T0094075

ROBOTS *and the* PEOPLE WHO LOVE THEM

ALSO BY EVE HEROLD

Beyond Human
Stem Cell Wars

ROBOTS *and the* PEOPLE WHO LOVE THEM

HOLDING ON TO OUR HUMANITY IN AN AGE OF SOCIAL ROBOTS

EVE HEROLD

ST. MARTIN'S PRESS
NEW YORK

First published in the United States by St. Martin's Press,
an imprint of St. Martin's Publishing Group

www.stmartins.com

Designed by Devan Norman

Portions of chapter 8 appeared on Leapsmag.com on December 31, 2018, and appear with permission of LeapsMag, now Leaps.org.

The Library of Congress Cataloging-in-Publication Data
is available upon request.

ISBN 978-1-250-12220-9 (hardcover)
ISBN 978-1-250-12221-6 (ebook)

Our books may be purchased in bulk for promotional, educational, or business use. Please contact your local bookseller or the Macmillan Corporate and Premium Sales Department at 1-800-221-7945, extension 5442, or by email at MacmillanSpecialMarkets@macmillan.com.

First Edition: 2024

10 9 8 7 6 5 4 3 2 1

CONTENTS

ROBOTS *and the* PEOPLE WHO LOVE THEM

INTRODUCTION

As a child during the 1960s, I was crazy about the futuristic cartoon *The Jetsons*. I swallowed the silly Jetsonian universe of flying cars, robot housekeepers, and moving sidewalks (perhaps the first Jetson-world innovation to actually come true) hook, line, and sinker. One of my fascinations centered on the idea of personal jet packs that would enable humans to fly. I thought that surely in a few years' time, I would have my own jet pack and even dreamed that someday we'd have talking dogs like the Jetsons' family dog, Astro, who would wear dog-to-human language translators on their collars.

While, like millions of other children, I was dazzled by the supposed technology of the Jetson world, there was always an underlying theme that I found reassuring, and that was the general incompetence of the technophobic George Jetson. Thankfully, the Jetson world, as replete as it was with all manner of advanced technology, was just as subject to common human foibles as any previous age.

George's wife, Jane, was a shopaholic, and George complained about his three-hour workdays. Rosie, the Jetsons' housekeeping robot, was prone to subversive wisecracks and had an authoritarian way with the children. The wizard-like home technologies that seemed to work so well for everyone else inevitably malfunctioned when George tried to use them, and who can forget his plaintive cry of "Jane, stop this crazy thing!" when the midair dog-walking treadmill went haywire? I decided that, if George Jetson could muddle through his days in our high-tech future, I had a chance of living in such a world.

One of my favorite episodes featured the introduction of Uniblab the robot at George's office at Spacely Space Sprockets. George's insecurity about being outmaneuvered by the robot is exacerbated when his boss, Mr. Spacely, makes no secret of his preference for Uniblab over all his human workers. According to Spacely, Uniblab is a marvel of efficiency and can do no wrong.

But Uniblab, far from being "just a machine," turns out to be a malicious, backstabbing coworker who quickly manages to get George fired. In addition to his sneakiness, Uniblab has a compulsion to trick humans out of their money by sucking them into gambling schemes that they can never win. Behind Spacely's back, Uniblab is a casino on wheels and a dishonest one to boot.

Eventually, George is able to unmask the robot by spiking his daily dose of lubricant with a secret ingredient that gets him drunk and causes him to act out disastrously in front of Mr. Spacely and the whole board of directors. Apparently, the brilliant robot is just as flawed, if not more so, than a run-of-the-mill human, and George's lazy, incompetent status quo is restored. More importantly, human nature, as flawed as it is, prevails over even the most advanced technology.

Science fiction, and even "pure" science writing that attempts to envision the future, often says much more about the time it was written than it says about the future world. George Orwell's dystopian novel *1984* was a reaction to the totalitarian societies created by Nazism and fascism around the time that Orwell wrote his book. The book was in fact an insightful reaction to real events that took the totalitarian nightmares of the mid-twentieth century to a logical extreme. However, once the year 1984 actually rolled around, the world, especially Europe, had changed dramatically. While 1984 looked quite different from the world of Orwell's novel, the book still stands as a brilliant illumination of the authoritarian mindset and its power, coupled with technology, to enslave humanity.

Of necessity, nonfiction books must hew more closely to the facts, even when speculating about the future. But books like this one can easily go off the rails when they project current values and conditions on a future world where a huge number of variables can cause society to change direction. There's no escape from extrapolating from the past and present to try to glimpse a bit of the future, and as such, our projected future world will greatly resemble the present. The only appropriate attitude of any writer who hopes to share some vision of the future is one of humility.

One of the central attributes of humans that will have a bearing on our relationship to robots is our status as social animals. We're easily the most gregarious creatures in the food chain, and this is already powerfully shaping our relationship to technology. Just thirty years ago, it would have been impossible to imagine how deeply embedded we would become with technology. Also unanticipated was the power of social media to engage, preoccupy, inform, mislead, and even manipulate us. In fact, the social aspects of our online activity have exploded to the point that millennials and Generation Xers now get more of their news from social media than from newspapers or television. We can debate whether this is a good thing, but as the best thinkers in the field suggest, technology that interacts with us in a social way, as robots are being designed to do, may be well-nigh irresistible.

In labs and factories across the industrialized world, social robots are being developed that give us an unprecedented opportunity to create a new species—an artificial one—in our own image. By *in our own image*, I mean not just that these robots will be in increasingly humanoid form but that they are also starting to resemble us in behavior. This artificial species will help liberate us from traditional labors and will provide a wide variety of valuable services. It will also tap into our emotional vulnerabilities and could even create a few new vices. Words used to describe it will include not only *social* but *emotional* and *ethical*. No matter how hard we try to conceive of social robots as mere tools, our lives will become deeply enmeshed with them, and over time, we'll build relationships with them that will challenge us to redefine what a relationship can be.

Robot nannies, friends, therapists, caregivers, and lovers will usher us deeper into digital culture, where new meanings about life and personhood will be born. These robots won't just be about cold, hard technology or even about the many labors they'll perform. Robots designed to learn from reading and understanding our words, expressions, and movements will be able to cater to our needs in ways that no human ever could. Because social robots are interactive, much of their role will be defined by the nontechnical things we humans bring to the relationship—emotions like affection, longing, and a desire for the magical in our lives. But there are also darker emotions like selfishness, dependency, and violence, which may find expression in how we treat our robots. What the robots bring out in *us* may be even more important that what they bring to the equation.

I pay special attention in this book to the desire for the magical and its deep-rootedness in childhood. We can all remember, if only fleetingly, a time when through our eyes the world was infused with wonder. Something about robots touches that inner child in us. We find them both intimately familiar and unfathomably strange. As children playing with dolls, teddy bears, action figures, stuffed animals, and simple gadgets, we naturally anthropomorphized almost everything we interacted with, and humanoid robots will tap into this tendency like nothing else. In one unbroken psychological continuum, we go from dolls to animated cartoon characters to windup gadgets to increasingly realistic robots.

The ability of these robots to read and react to human emotions and even to feign emotions of their own is something that no other machine has been able to do. As robots become evermore sophisticated, we may have to keep reminding ourselves that no actual magic is involved.

It's an open question of whether robots will ever possess consciousness, an inner life to which we can make a true connection. This is one of the most intriguing questions in all of science. Robots are already feigning consciousness to an uncanny degree, yet much of the "robot mind" is still not understood even by the people who design their algorithms and endow them with a "mind." In other words, no one today has a perfect understanding of what goes on in the brains of robots, and that which is not understood can easily be taken for magic. Imagination floods in where logic falls short.

To make matters even more complicated, human beings are easily beguiled, especially when our emotions and our social natures are engaged. We may have less control over our robot relationships than we think.

One of the most difficult questions that roboticists are frequently asked is when highly advanced AI will become a reality. There are many opinions on the subject, and I've resisted the urge to get deeply involved in speculation about what even the most knowledgeable experts disagree on. But I assume that highly advanced AI will become a reality and that working to anticipate its inevitability is not only worthwhile but necessary.

Rather than focusing on far-future predictions of what the world will look like in the next hundred years, I've tried to stay grounded in technologies and issues that already present themselves. The early-stage social robots already in existence are enticing enough to create complex relationships with people. These robots are still fairly simple in their range of behaviors. But I've learned in the course of writing this book that the reactions they elicit in humans are anything but simple.

Perhaps even more important than the attributes of social robots are the complex and emotionally laden issues we project onto them. Robots easily become a kind of Rorschach test that reflects back to us our own inner worlds. There is evidence to suggest that socially interactive robots can be therapeutic, not just for the physically impaired but for children and adults with cognitive and emotional issues, such as autism, ADHD, depression, and dementia. As I'll explain, it doesn't take a particularly able robot to elicit projection or to bring out the very best and the very worst in humanity.

One danger of social robots is that our behavior toward them will not involve the same consequences that our behavior toward humans entails. The unhindered habits we develop in interacting with our robots could easily carry over into how we treat other people and animals, where there are real-life consequences. And given the special role that the family robot will play in caring for, educating, and entertaining children, this consequence-free behavior could develop quite early and be reinforced by the robot's inability to push back.

Another problem is that, despite a significant body of research into human-robot interaction, few experts seem to be asking the question of whether socially interactive robots will have a net positive effect on society. Fewer still question the idea that social robots will develop increasingly complex and humanlike traits, attributes that will make them ever more engaging, more convincing in their quest to be lifelike. Very little research asks whether we *should* have social robots; the vast majority is focused on how to make robots more appealing in order to better facilitate their acceptance.

Arguably, it's not the job of roboticists to ask these types of questions—such discussions fit better in the realms of philosophy, psychology, and social studies. I've given the most attention to these areas because, as the science of building social robots rapidly progresses, we might employ a healthy sense of skepticism when the seeming magic of robots threatens to obliterate common sense. At the same time, I confess that the charm of social robots is considerable and extremely hard to resist. That's exactly why we need to get a clear handle not just on what they are but on what they are not.

The technology behind the development of social robots is moving forward at such a rapid rate that in a few years' time, the innovations covered here will have been far surpassed. Yet human nature changes at a glacial pace if at all. Because of this, I've endeavored to center the discussions in this book not so much on the capabilities of the robots but on the human-robot relationship. And here is where I think the enduring qualities of humanity, spurred by new types of relationship, may give rise to something truly novel.

We have no cultural conventions to guide us in how to respond to robots; we must develop them as we go along. As time goes by, robots will make their own contributions to human culture and may take us in directions that currently no one can anticipate. Much depends on whether humanoid robots will be endowed with advanced artificial intelligence or whether we will, by design, decide to limit their abilities. If past history is any guide, if the technology exists, we will embrace it.

Are we up to the challenge of keeping robotic relationships in perspective and making sure that they don't replace our interactions with real humans and even companion animals? Will real social, ethical, and technical skills atrophy as relying on our robots becomes the path of least resistance? However we envision answers to these questions, one thing is certain: Our very concept of relationship is about to be challenged, questioned, and ultimately redefined. Our dependence on technology is about to reach an unprecedented level, and our understanding of ourselves is likely to be changed beyond our wildest dreams.

1

THEY'RE HERE

In the early 1900s, the artist Oskar Kokoschka was the blond, rakishly handsome enfant terrible of the Viennese art scene. Known as much for his colorful personal life as for his expressionist paintings, the press referred to him as "the wildest beast of all."

One of the most fateful encounters of his life came in 1912 when he met Alma Mahler, the beautiful, dark-haired widow of the famous composer Gustav Mahler. The recently bereaved Alma was one of the most desired women in all of Vienna, and the two of them plunged headfirst into a tumultuous love affair. Alma became Oskar's artistic muse, and when they weren't making love, he was compulsively drawing and painting her.

Soon, however, Oskar's passion turned to obsession and, for Alma, his jealousy became a prison. After three stormy years,

Alma broke off the affair, telling Oskar that she was afraid of becoming too overcome by passion. He was devastated and even more distraught to learn that she had aborted his baby, an act that he took as the ultimate rejection.

To try to win Alma's admiration (and perhaps her sympathy), Oskar volunteered as an Austrian cavalryman in World War I. He was soon seriously wounded and was twice diagnosed with shell shock, an antiquated term for post-traumatic stress. While he was recovering in a hospital in Dresden, doctors decided that he was mentally unstable and had him discharged from the army. Oskar returned to Vienna only to discover that, rather than welcoming him with open arms, as he had hoped, Alma had taken up with another man. He was plunged into despair.

The year was 1918, and Oskar decided that if he couldn't have Alma, he would have a life-size doll made in her likeness. He hired the Munich doll maker Hermine Moos and provided her with numerous detailed sketches and a life-size oil painting of Alma. He was exacting, even obsessive in his instructions to Moos, hoping for the most lifelike copy of Alma possible. He wrote to the doll maker:

> Yesterday I sent a life-size drawing of my beloved and I ask you to copy this most carefully and to transform it into reality. Pay special attention to the dimensions of the head and neck, to the ribcage, the rump and the limbs. And take to heart the contours of the body, e.g., the line of the neck to the back, the curve of the belly. Please permit my sense of touch to take pleasure in those places

> where layers of fat or muscle suddenly give way to a sinewy covering of skin . . . The point of all this for me is an experience which I must be able to embrace!

Later, he asked, "Can the mouth be opened? Are there teeth and a tongue inside? I hope so."[1] As the months of making the doll progressed, it became clear that Oskar was not just hoping for a reminder of Alma; he sought to replace her altogether. In anticipation of his doll, and perhaps to make Alma jealous, Kokoschka had his maid spread rumors throughout Vienna that he had a new woman in his life, and curiosity about who this new woman could be abounded.

When the doll finally arrived, Oskar took his fetish to new levels. He took the doll for long carriage rides and rented a box at the Vienna opera house, where he seated her next to him. He sat with her at sidewalk cafés, where he carried on conversations with her. He even had his maid attend to her, dressing her in fine clothes and treating her like the lady of the house. "See," he seemed to be saying to Alma, "I no longer need you because a beautiful doll can take your place!" He painted hundreds of portraits of the doll, over one hundred of which featured the two of them together, seemingly as a happy couple.

Kokoschka's doll dissolved the boundary between life and art, allowing him to recede into a fantasy world that subsumed the reality of his love life. She also gave him an object on which to act out his unresolved feelings toward the real Alma. Eventually, he managed to do to the doll what he would have liked to do to his lost love. The end of the relationship was just as florid and dramatic as its commencement. He wrote:

> I engaged a chamber orchestra from the Opera. The musicians, in formal dress, played in the garden, seated in a Baroque fountain whose waters cooled the warm evening air. A Venetian courtesan, famed for her beauty and wearing a very low-necked dress, insisted on seeing the Silent Woman face to face, supposing her to be a rival. She must have felt like a cat trying to catch a butterfly through a window-pane; she simply could not understand. Reserl paraded the doll as if at a fashion show; the courtesan asked whether I slept with the doll, and whether it looked like anyone I had been in love with . . . In the course of the party the doll lost its head and was doused in red wine. We were all drunk.[2]

The next morning, Kokoschka's doll was found naked and beheaded in his garden. Apparently having worked out his issues with Alma, Kokoschka no longer needed her.

His bizarre acting out was extreme, but as often is the case with eccentricity, Kokoschka's need to project his emotional issues onto an inanimate object is only a matter of degree, not kind. The doll allowed him to channel his obsessive feelings in a relatively harmless way; better that he beheaded a doll rather than murder his real-life love. Ultimately, the doll could not return his affections. It couldn't speak or respond to his caresses or even feign emotion. Yet it played an integral part in his effort to resolve the issues he had yet to work out with the real Alma. Kokoschka went on to live a long, characteristically eccentric life and died in 1980.

A hundred years later, we have Pepper, the social robot. Pepper is not an erotic symbol but is rather in the form of an innocent, rather needy child. Made of metal and plastic, he is just under three feet tall and looks as much like a cartoon character as a robot. His large, round eyes peer out of a round head, giving him the air of an inquisitive toddler. When you speak to him, his eyes gaze up at your face, and he follows your movements with his head.

Interacting with Pepper is effortless because he takes the initiative. All you have to do is move and speak, and Pepper will start a conversation with you in his childlike voice. But Pepper can do far more than a mere doll. Not only does he respond to your speech, he "reads" your expressions to ascertain your emotional state and responds with what his makers deem is appropriate behavior. For example, if Pepper notices that you look sad, he'll try to cheer you up by playing your favorite song. And if you don't feel like hearing your favorite song, just tell him so in the natural language you would use to address a real person. The music will stop, but Pepper isn't likely to give up. He's on a mission, it seems, to make you feel better.

Created by the French robotics company Aldebaran, which is now owned by SoftBank, the Japanese telecommunications firm, Pepper is one of the first humanoid robots designed specifically for companionship. As rudimentary as he is, he's light-years more advanced than Kokoschka's inanimate doll.

He "learns" from interacting with humans and is able to develop a unique relationship with them by remembering their tastes, desires, and even their moods. His makers describe him as "engaging and friendly." Some people might find the chatty robot, at least some of the time, a bit too friendly. However, that's easily corrected without the need for any technical skills whatsoever. All you have to do to program Pepper is tell him what you want, and he'll respond accordingly and even remember your words for future reference.

But Pepper is only part of the equation; human nature fills in the rest. Regardless of our individual needs, and thanks to our hardwired emotional natures, most of the time we're tempted to interact with him as though he were human. Research has shown that, like it or not, robots like Pepper can push our emotional buttons and elicit responses from us that defy our rational awareness that they're not actually living things.

Unlike his industrial cousins, Pepper is a consumer robot meant to live with people in their homes. Unfortunately, he doesn't do windows or carpets or dishes. He's designed solely to be a companion. Currently priced at $1,900, he's relatively affordable, and so far each month, when one thousand Peppers go on sale in Japan, they sell out within minutes. Pepper's creator, roboticist Kaname Hayashi, is unabashed in his hope that the talkative robot may be able to banish loneliness. "We all feel lonely," Hayashi has said. "We lie if we say we don't."[3] Pepper not only simulates emotions and behaves as though he has empathy for our feelings, he elicits empathy in us, a phenomenon that evokes both delight and trepidation.

Popular culture has long been replete with books, stories, comics, movies, and games whose central theme is one of robots usurping humans and taking over the world. It's hard to say if Pepper, in particular, will take over the world, but there's no doubt that some version of a companion robot will be coming soon to homes throughout the industrialized world.

For those who doubt that robots are about to become ubiquitous, consider that we're already surrounded by them. Not only are robots used widely in manufacturing and shipping, they go inside volcanoes testing for toxic gases, clean up radioactive materials after nuclear accidents, sweep for undersea mines, collect intelligence and test for explosive devices for the military, and explore other planets. Robots developed for oceanographic research descend into depths where no human could go, and some are designed to swim like fish or swarm like insects. They help doctors make diagnoses by almost instantaneously analyzing huge amounts of medical information, determine people's eligibility for Medicare and Medicaid, and even perform surgery. The Japanese have launched (literally) a robot that can repair the International Space Station, and it's anticipated that robots will soon build research stations on other planets, supervised by astronauts in orbiting space stations. In addition to the work they'll do, these robots will also provide entertainment and companionship to astronauts on long journeys into space. Despite a lack of fanfare to announce their proliferation, robots are already an integral part of modern life.

Police organizations throughout the world are already employing robots to do especially dangerous jobs or to go where police officers can't go. The Ohio state police are using a six-wheeled robot to explore tight spaces in search of bombs, and in India, police riot-control drones can watch crowd behavior and, if needed, fire pepper spray, paint pellets, or tear gas. Israeli counterterrorism forces have a land-based rover that packs a 9 mm Glock pistol and can enter a house, maneuver over obstacles, and even climb stairs, all the while transmitting information to operators through cameras and a two-way radio. Japanese police forces have a flying drone that can locate and shoot down hostile drones, and the Greek coast guard has a robotic flotation device that can rescue drowning refugees who are struggling to cross the Mediterranean Sea. In the Democratic Republic of the Congo, more humanoid-looking robotic traffic cops direct traffic at busy intersections, and at a prison in South Korea, a five-foot-tall robot patrols hallways and uses pattern-recognition algorithms to detect problem behaviors on the part of prisoners.[4]

By adding more and more capabilities that allow them to interact with humans, roboticists are greatly expanding the roles robots can play in everyday life. The government of Switzerland is testing a small, six-wheeled robot that can read human handwriting and negotiate outdoor environments well enough to deliver the mail.[5]

Interactive robots now help travelers at several airports across the world. In the Geneva Airport, a boxy, autonomous robot named Leo can greet you, check in your bags and deliver them to the correct handling area, give you up-to-the-minute information about your flight and boarding gate, and direct you to locations like the nearest restroom or ATM.[6] At Amsterdam Airport Schiphol, a freewheeling robot with a slightly more humanoid appearance named Spencer interacts with passengers by helping them navigate the airport. Spencer can communicate in several languages and understands human behavior well enough to know how to navigate around people even in a crowded terminal.[7] At Narita International Airport in Japan, Honda's walking, talking robot Asimo greets weary international travelers before they line up for customs and usually manages to put a smile on their faces. Sometimes Asimo's greeting involves jumping and kicking a soccer ball, a performance that inevitably draws applause. Travelers who venture toward the Bank of Tokyo's branch at the airport's first terminal will encounter NAO, a humanoid robot similar to Pepper, who provides currency exchange rates and guidance to airport facilities in Japanese, Chinese, and English. The two-foot-high robot, who stands at eye level on a counter, "blinks" its eyes and makes lifelike gestures as it engages in conversation with curious people of all ages and nationalities.[8]

The most important capability of airport robots is to recognize and process language and respond appropriately to humans. In this respect, one might consider them to be not much more than interesting novelties, but with each passing day, roboticists, using specially written algorithms, are adding more advanced, humanlike capabilities.

Robots, some embodied and some not, are taking over human jobs in a wide array of professions. Starting in September 2016, the home improvement store Lowe's began testing the LoweBot customer assistance robot in eleven of its San Francisco stores. The robot scans inventory and leads customers to whatever tool, appliance, or gadget they're looking for simply by being asked. For those who feel uncomfortable talking to a robot, the LoweBot also has a touch screen for communication. Robots like the LoweBot are likely to become a common aspect of the shopping experience as they replace salespeople and cashiers. Officials for Lowe's insist that their robot won't lead to layoffs for their human customer service providers—it will free them of the most repetitive tasks so they can spend more one-on-one time with customers.[9]

But robots are able to do considerably more advanced tasks than just walking, talking, and processing language. Their abilities are growing at a rapid pace, and although we already depend on them in a host of online capacities, in many cases, the interaction is so seamless that we're not even aware that a robot has played a role.

Many people are unaware that twenty-three million Twitter users are automated bots. If you follow a Twitter user named Olivia Taters, you're following a tweet-generating bot that was designed to speak in the voice of a typical teenager, created by Rob Dubbin, a writer for the comedy show *The Colbert Report*. Dubbin also created a bot designed to churn out the kind of tweets sent by conservative news organizations, called Real Human Praise. Both of his brainchildren have thousands of followers, many of whom think they're following the Twitterized stream of consciousness of real people.[10]

This phenomenon is a real problem for Twitter, whose stock value is based on its audience reach as an advertising platform. Potential advertisers want to feel certain that they're reaching real people, and when they lose confidence in Twitter's real audience reach, they decline to buy advertising and the company's stock value falls. This has already happened more than once, when news about the huge number of Twitterbots came out. When Elon Musk was in negotiations to buy Twitter, the number of Twitterbots became such an issue that it almost scuttled the deal. But equally important, people who use Twitter are emotionally engaged because they believe they're following and communicating with other people. The fact that automated bots are able to mimic the communication patterns of real humans is only the tip of the iceberg in human-robot interaction (HRI). Robots are already fooling us in myriad ways.

Robots used to be confined to replacing human workers in tasks that are dirty, dangerous, and repetitive. Although they still fill that role in many industries, they're starting to fool us into thinking that we're dealing with humans in a growing array of capacities. Some bots are now doing things that until very recently we thought only humans could do, tricking even the most technically savvy among us into thinking they're dealing with a real person.

In 2016, Georgia Institute of Technology College of Computing professor Ashok Goel added a new teaching assistant named Jill Watson to help with his course in knowledge-based artificial intelligence. The online course is required for a master's degree in computer science, and each year, about three hundred students post roughly ten thousand messages in online forums, mostly questions about the material being studied—far too many for Dr. Goel and his several teaching assistants to respond to. Goel added the new teaching assistant to answer student questions and provide them with feedback to keep them on the right track.

After being trained by reviewing about forty thousand past questions that students had actually asked, the new TA was soon able to answer questions with 97 percent accuracy, and Goel put her to work fielding student questions online. The students' response was uniformly positive. These were postgraduate computer science students, and what none of them realized was that Jill Watson was actually an online bot whose interaction was so seamless that it was virtually indistinguishable from a real teaching assistant. Goel's bot is currently handling 40 percent of his students' online questions.[11]

The public has long been concerned about robots taking over jobs involving manual labor and thereby displacing human workers. This concern is well founded; robots have indeed taken over a large number of manufacturing jobs and are projected to take over many more, if not most of them, in the next few decades. What few realize is that they're getting more and more adept at performing at near-human level in many white-collar jobs.

Fields like law and journalism are already being affected. In 2014, the highly respected news agency the Associated Press (AP) started publishing routine stories with no byline attached—that is, no human byline. The stories are being generated by "journalist" bots and read just like typical AP stories, only missing a subjective point of view. The fact is millions of articles by bots are being published each year not just by the AP but by large companies like Comcast and Yahoo. These publishers rely on bots produced by an article-generating company in Durham, North Carolina, called Automated Insights, and the vast majority of readers take for granted that they're reading the work of traditional journalists. Not only have the bots mastered tone and style well enough to fool readers into thinking they're human, they can generate up to two thousand articles per second.[12] Journalist bots are especially useful in fields that require the crunching of large amounts of data, such as finance and sports reporting, and in 2015, Automated Insights produced 1.5 billion "narratives" used in published news accounts.

More recently, the chatbot ChatGPT, created by the company OpenAI, has been covered extensively in the news for its impressive ability to scan the internet for information and cobble together essays that sound, to many people, as though they've been written by real people. Many observers are troubled by the possibility of this technology eliminating a plethora of white-collar jobs. A spirited debate has emerged over the accuracy (not 100 percent) and even the sometimes aggressive and malevolent content ChatGPT generates in conversations with people. I'll explore such phenomena in a later chapter.

Bots are quickly going beyond the written word and may soon be generating video journalism as well. The basic technology was unveiled in 2016 by two Georgia Tech computer science students, Daniel Castro and Vinay Bettadapura, and it utilized the elements of video editing that could easily fool one into thinking that a human editor was involved. The students wrote an algorithm that turned twenty-six hours of raw vacation video footage into a thirty-eight-second highlight video in just three hours.

The bot analyzed the raw footage for location, image composition, symmetry, and color vibrancy, scoring each frame. Then it selected the footage with the highest scores and even selected the most picturesque content. And the program, just like a real editor, can create videos that are highly reflective of the user's tastes and desires. Bettadapura said, "We can tweak the weights in our algorithm based on the user's aesthetic preferences. By incorporating facial recognition, we can further adapt the system to generate highlights that include people the user cares about."[13] This algorithm could easily be incorporated into the abilities of an embodied household robot and make the production of highly aesthetic, meaningful videos that highlight important events in an individual's or a family's life—such as a birthday party—as well as take on much of the workload for professional videographers.

Inventors keep pushing the limits of what machine intelligence can do, taking their inspiration from human beings. The direction of robot research is making robots do more and more things that we have always believed could only be done by humans.

Google Brain, Google's artificial intelligence (AI) project, is trying to teach computers to be creative. In 2016, an initiative called the Magenta project fed four musical notes to a computer and "primed" it to write a piece of music. The result was a ninety-second song using the sound of a piano set to a rapid beat. The song is fairly repetitive but is not unlike any number of electronic tunes that make up the components of popular songs.[14] The piece may not have anywhere near the complexity or nuance that a human composer could create, but one can easily see how such bot-generated tunes could be used in sampling by human DJs and songwriters, augmenting music as we know it. As the technology advances, it's no longer so certain that humans will continue to be the only beings capable of producing music.

Google's ambitions to teach machines to be creative doesn't end with the Magenta project. It has created DeepDream, a program that sorts through, alters, and remixes photographs to create nightmarish, surreal works of art that are mesmerizing. Google uses a system mimicking human neural networks to create images that critics describe as hallucinogenic. Looking at the pictures, it's hard to believe they weren't done by an extremely imaginative artist.[15]

Adding "creative" algorithms to computer programs that can be incorporated into robots adds to the eerie impression that robots could have a bona fide inner life. This is especially powerful if such attributes are given to robots that are socially interactive because it invites us to have empathy for them and for their assumed inner lives. In fact, even without such humanlike cues, we're already subject to feeling empathy for even simple robots.

As mentioned, people are hardwired to feel empathy for machines that display even the most rudimentary behavior, and we may even be hardwired to anthropomorphize, or project human attributes onto, the simplest of robots.

Floorance, Darth Roomba, Sarah, Alex, and Joe. These are some of the names people have given their Roombas, the simple, disklike robots that wander their homes vacuuming carpets and floors. When researchers at the Georgia Institute of Technology surveyed 379 Roomba owners to profile how early adopters of robot technology feel about their robots, they tapped into a cornucopia of anthropomorphic feelings about the Roomba.

More than half of the owners ascribed a gender to their Roombas, and about a third of them named them. A large percentage of them ascribed a personality to the humble appliance and talked to it, even praising it for doing a good job. Forty-three of them bought a costume for their Roombas and dressed them up as fictional characters.[16] All this and the Roomba doesn't come close to the level of sophistication of a social robot that can read and express emotions. The Georgia Tech study only highlights what psychologists already recognize: interactive robots are not really about robots—they're mostly about us and the complicated emotions we bring to the human-robot interaction.

As household robots become more humanoid and realistic, we're likely to project far more humanlike traits on them because of our natural tendency to anthropomorphize. Anthropomorphism is deeply ingrained in human psychology and has been a prominent feature of religion and culture since the dawn of recorded history. The word *anthropomorphism* combines the Greek words *anthropos*, meaning "human," and *morphe*, meaning "shape" or "form." From believing in legends like the man in the moon to seeing faces in clouds and hearing voices in the wind, anthropomorphism starts very early in childhood and persists throughout life. Adults are better than children at refuting and rationalizing the tendency, but anthropomorphizing animals, toys, and even mechanical devices such as cars and computers is well-nigh universal, and cultures throughout the world reflect and reinforce it.

Ancient mythologies and religions frequently depicted gods as having human forms and personality traits. Statuary of anthropomorphic gods and goddesses have been found in archeological sites throughout the world, including those from ancient Egypt, Greece, and Rome; Mayan and Aztec carvings; African masks and statues; and carvings in Hindu temples. The Greek gods and goddesses were thought to display not only human form but human emotions like love, pity, jealousy, and anger. They fell in love, engaged in petty rivalries, and wreaked havoc on humanity because of their passing whims. But their humanlike traits served an important purpose—to render them understandable to the struggling humans who had to endure the vicissitudes of a sometimes cruel and indifferent world.

The Scottish philosopher David Hume believed that we apply human traits to nonhuman agents in order to understand a mysterious, often confusing world. Sigmund Freud described the tendency as a way of making a threatening world seem more friendly and manageable.[17] In addition to filling an emotional need, anthropomorphism fills in the unknown with what is presumably known: human nature and personality.

From the first cave paintings to the most elaborate animated films today, anthropomorphism has been one of the most effective devices in storytelling and literature, especially the kind aimed at children Written around 700 BC, Hesiod's story "The Hawk and the Nightingale," which depicted highly intelligent, talking animals, preceded Aesop's fables by centuries. Aesop's tales of wily, humanized animals are still loved by children and adults alike, teaching wisdom to children through the adventures of highly anthropomorphized animals by embodying traits that we encounter daily with humans.

For centuries, fairy tales have featured talking animals and personified elements such as Zephyrus, the west wind, which carried the nymph Psyche away. The personification of gods, animals, and natural elements adds to a sense of magic and enchantment that is readily embraced by both children and adults. Robots that look and act even somewhat human click into the pattern with enough ease that we're scarcely aware that we're anthropomorphizing at all.

In more recent times, Lewis Carroll enchanted whole generations through his white rabbit and animated playing cards in *Alice's Adventures in Wonderland* (1865). Carlo Collodi elicited a sense of magic and delight through his animated wooden puppet, Pinocchio, who comes to life and embarks on a quest to become a real boy in *The Adventures of Pinocchio* (1883). And Rudyard Kipling's jungle in *The Jungle Book* (1894) teems with the colorful antics of talking, wise, and helpful animals. In the tales of Beatrix Potter, animals like Peter Rabbit resemble naughty and mischievous children who get themselves into, then must find their way out of, all kinds of predicaments. Children and adults alike have reveled in the enchanted worlds of *The Wind in the Willows* (1908) by Kenneth Grahame, *The Lion, the Witch, and the Wardrobe* (1950) by C. S. Lewis, and *Winnie-the-Pooh* (1926) by A. A. Milne, tales involving intelligent animals whose motives and aspirations appear decidedly human. Literary anthropomorphism even took a political turn in George Orwell's 1945 novel, *Animal Farm*, in which the animals act out an allegory based on the Stalinist era of the Soviet Union. But the undisputed kings of anthropomorphism hail from Hollywood.

In the contemporary era, including during the childhoods of most adults today, anthropomorphism has reached its apex through the films and television shows of Disney, *Looney Tunes*, and Pixar. Disney's Mickey Mouse, Donald Duck, and even an anthropomorphized magic carpet (from the *Aladdin* films) are beloved throughout the world, as are Bugs Bunny, Daffy Duck, and Porky Pig from *Looney Tunes*. Equally recognizable are Pixar's animated toys from *Toy Story*, who magically come alive when no humans are looking, and the anthropomorphized automobiles in the films *Cars* (2006), *Cars 2* (2011), and *Cars 3* (2017).

Contemporary culture, especially child culture, is steeped in tales of personified animals and artifacts. The recent trend in roboticized toys like the robot dog Aibo simply pick up on where TV shows, films, and video games leave off, appealing to our ancient tendency to ascribe human feelings and intentions to all sorts of nonhuman objects. Humanoid robots that speak, process language, and feign emotions tap directly into not only culture but into the hardwired patterns of the human psyche.

The reasons they appeal to us go as deep as our innermost conscious and unconscious desires for understanding our world, mastering our environment, and, most of all, our need for social connection.

Although anthropomorphism is exceptionally common, there was not a great deal of research on the phenomenon until 2007, when psychologists Nicholas Epley, Adam Waytz, and John Cacioppo at the University of Chicago published a paper in which they identified three deep-seated drives that make humanizing the nonhuman almost irresistible. One of those factors, which they call "elicited agent knowledge," springs from the fact that we must anchor our understanding of the unknown in what *is* known—namely, human character traits, emotions, and desires.

From earliest childhood, as we seek to understand the world, we engage in a process of inductive reasoning—starting with the specific (ourselves) and extrapolating from the specific to the general. We assume that what we know to be true of ourselves can be applied to the world at large. The child reasons that if she is afraid of the dark, it follows that her dolls and stuffed animals are also afraid of the dark. This may be an inaccurate assumption, but the process is inescapable, as the human child must have a starting point when seeking to understand the world. Later, the child may revise her understanding based on more experience, but it was always necessary to have a starting point.

The second psychological drive identified by Epley et al. is what they call "effectance," or the need to interact effectively with one's environment. Projecting human traits onto toys, animals, and machines helps to alleviate the pain of uncertainty and replace it with a sense of understanding. If I project human traits and feelings onto a robot, I believe I can better predict its behavior and have a greater feeling of control.

As adults, we get better at later countering our inaccurate assumptions about those things to which we first ascribed human traits, but the anthropomorphism was still essential as a starting point. And even though we can later revise our understanding of things like robots and toys, we may not want to relinquish the flights of imagination that led us to anthropomorphize them in the first place. Thinking that a robot has human feelings and motivations provides us with a useful framework for predicting future behavior, and it satisfies what may be the most compelling drive of all—the need for social connection.

Epley and colleagues place the need for social connection at the center of human drives, comparable in importance to hunger and thirst.[18] They write that "anthropomorphism enables satisfaction of this need by enabling a perceived humanlike connection with nonhuman agents."[19] The need for social connection and social approval is ingrained in humans from infancy onward, and it is so powerful that we automatically seek it from both human and nonhuman agents. We make connections with other humans by first attributing to them our own feelings and traits and later revising or supplementing our understanding with more factual information.

Social connectedness is so crucial that it's literally a matter of life or death. The link between health and social connectedness has been well established, and the effect persists throughout life. Loneliness, or a lack of social connection, is worse for one's health than smoking, high blood pressure, or obesity.[20] Studies have shown that social relationships have profound short- and long-term effects on one's life and even have a cumulative effect on our mental and physical health over a lifetime.

Adults with coronary artery disease who are socially isolated have a risk of cardiac death 2.4 times higher than that of their more socially connected peers. Loneliness has been linked to not only heart disease but to everything from cancer to slow wound healing to impaired immune function.[21]

Feelings of social connectedness, the sense of being loved, cared for, and understood, tend to inoculate us from both mental and general health disorders from infancy to old age. Pet animals as well as people can provide the many benefits of social connectedness. The unconditional love and acceptance of pets is one reason why having animal companions proves so therapeutic, and our anthropomorphizing of them plays an integral part in why the relationships are so powerful.

Epley and colleagues note that those who *lack* a sense of connectedness have an even stronger drive to relieve their loneliness through anthropomorphizing others, whether human or nonhuman. The need to make sense of, gain mastery over, and be socially connected to agents in our environment easily leads us to develop empathy, the very glue that holds society together.

A wealth of research is bearing this out. Recently, the findings of a collaboration among Japanese computer engineers and psychologists were released. The research team had performed electroencephalographys (EEGs) on fifteen people as they watched videos of humans and robots in perceived pain. It turned out that the brains of the human subjects reacted almost the same way to the robots' pain as they did to perceived pain in a human.[22] The neurophysiological evidence only reinforced what roboticists have long observed: people feel empathy with robots that is similar to the empathy they feel for other people. While this improves the likelihood that we'll integrate social robots into our lives, it unleashes a plethora of issues related to the relationships we develop with robots.

"Our entire civilization is based on empathy," says the University of Calgary computer scientist Ehud Sharlin. "Societies are built on the principle that other entities have emotions."[23] It doesn't take a particularly humanlike appearance for owners to attribute all sorts of human attributes to their robots and to respond to the illusion as if it were alive. It's a quality MIT psychologist Sherry Turkle calls our "social promiscuity," evidence that we humans are "cheap dates" when it comes to being seduced by robots.

Turkle has concluded through research with both children and adults that even simple robots become "powerful objects of psychological projection," used by humans to work out their emotional issues with other people. People of all ages quickly develop a need to nurture their robots and be nurtured by them. The illusion of nurturance, according to Turkle, could be the new "killer app" that will make social robots ubiquitous.

Turkle and others have written about a nascent "robotics culture" born in the interactions of humans and robots. Key to the development of this culture will be the suspension of disbelief, something we have been good at since childhood. It's likely that humans will become comfortable with relationships that are a blend of authenticity and artificiality. As technology advances and social robots become ever-more intelligent, our own culture will evolve new categories of relationship that incorporate robots and artificial intelligence.

In the next few years, more robots will become personal and social. "Personal, social, emotional, home robots," as they are termed in the robotics business, are envisioned as the next general-purpose technology. Like cell phones before them, the household robot will consolidate a wide range of functions, including communicating with the web, controlling multiple functions in a "smart" home, providing security surveillance and recording events, reading and responding to emails, posting updates on social media, playing music and movies, performing basic housekeeping functions, ordering food, reading children's stories to them, providing companionship for the lonely, and much more. They will teach, entertain, and amuse. By consolidating multiple technologies, personal robots will declutter our lives and become the main technological interface in our homes, controlled by simple voice commands.

While such versatility in any electronic device is attractive enough, robots like Pepper will soon make themselves indispensable because of their unique, highly personalized relationships with us. They can already read our words, facial expressions, gestures, and life histories and incorporate this information into their interactions with us. We will continuously train these robots simply by interacting with them. They'll "know" us and act accordingly, seeking to satisfy our every whim. But will such relationships be good for us in the long run (i.e., will they contribute to our flourishing as social beings who are highly interconnected with other beings), or will they stunt our social, emotional, and intellectual skills due to a lack of genuine relationships?

Another question is, because social robots will be so versatile with managing the various technologies that we depend on, will we lose skills that are necessary for maintaining independence and autonomy? We're already drifting away from understanding the sea of technology that supports our lives, and as robots become abler, we're likely to outsource an increasing number of functions to them.

The more able we are to rely on robots, the likelier we are to do so. The question is whether we'll become socially isolated slackers who trade real relationships and our self-reliance for dependence on machines that cater to our hardwired emotions and social tendencies.

2

OVERCOMING THE UNCANNY

Robots can be captivating and endearing. They can also be creepy, unsettling, and terrifying. Imagine waking up in the middle of the night to the presence of a humanlike figure standing silently in your bedroom. The figure starts to move, and you have no idea what it's going to do next. The figure is a robot, and there is no one else in the house with you. What goes through your head? It will do no good to have robots in our houses if we develop an unhealthy fear of them, and it's this barrier, planted deep in the human psyche, that engineers must cross if we're to bring them so intimately into our lives.

Now picture two Japanese men sitting side by side. They could easily pass for twins, identical down to the pores of their skin and the networks of veins on their hands. They have the same straight black hair, glasses, and taste for dark-colored clothes. They even share a facial expression, characterized by two small, vertical creases between the eyes, which gives them a rather brooding look. They speak in the same voice. The only difference: the figure on the left is Professor Hiroshi Ishiguro, a leading Japanese roboticist, and he created the figure of his robot doppelgänger, named Geminoid HI-1, the figure on the right.

When the highly lifelike robot speaks, his flexible, rubberized face is almost as animated as his maker's, but there's an eerie glitch. His words are ever-so-slightly unsynchronized with his lips, and suddenly the spell is broken. The robot subtly reveals his artificiality, we realize that the eyes appear fixed and glazed, and his brooding expression becomes creepy and menacing. He has fallen into what roboticists call *the uncanny valley*.

The uncanny valley was first described by the roboticist Masahiro Mori in 1970, when he illustrated that the human response to robots becomes more positive the closer a robot's appearance approaches that of a real human. However, once the robot passes a certain threshold for believability, the resemblance becomes unsettling and even scary, bringing to mind visions of corpses and zombies. Inevitably, the humanoid robot will reveal some subtle betrayal of reality, and what was formerly charming becomes simply macabre.[1]

Freud wrote about the experience of the uncanny as one in which the long familiar becomes strangely, eerily unfamiliar, or, put another way, when something that is strangely unfamiliar, like a robot, embodies aspects of the long familiar, like a human emotional expression.[2] Certain wires get crossed in the brain of the beholder, and the effect is one of anxiety, dread, and even fear.

Perhaps if we hadn't been so taken in by Geminoid HI-1's realism, we may not have reacted so strongly to the crack in his façade. Psychologists and sociologists have found that, for robots to be accepted into the most intimate parts of our lives as caregivers, teachers, companions, and assistants, they need to be realistic enough to convince us that they are like us. However, if realism is the goal, it has to be seamless. Otherwise, some subtle glitch will undoubtedly cast us into the uncanny valley. At the same time, for some types of robots, ultrarealism may be neither necessary nor desirable.

Dr. Ishiguro's robots are some of the most humanlike in the world of social robots, more accurately referred to as *androids*, and androids, according to many studies, are the most likely to elicit feelings of uncanniness. Since creating Geminoid HI-1, Ishiguro has designed two friendly-looking female androids, named Kodomoroid and Otonaroid, who are now employed as robot guides at Japan's National Museum of Emerging Science and Innovation. They look like pleasant and approachable young women and are remarkably lifelike in appearance.

But according to those who have interacted with them, these two robots haven't escaped the uncanny valley. In fact, their visual fidelity to real life only exacerbates the eerie effect when their speech becomes slightly unsynchronized or they make some arm movement that appears jerky and unnatural. Visitors to the museum are drawn to the androids, yet creeped out and repelled by their minor flaws. Engineers and roboticists are well aware of the effect, and it's widely recognized that for robots to be accepted into society, they will have to somehow avoid or counteract the uncanny valley.

There are several theories about the origins of the uncanny

valley, which apply not only to robots but to lifelike dolls and computer animations. The roots of the uncanny valley lie deep in human psychology and seem to persist to some degree across all nationalities and cultures, which suggests that the problem could become a real stumbling block to widespread robot acceptance. It seems the greater the effort to make the robot lifelike, the more pronounced the feelings of creepiness and revulsion when some discrepancy occurs.

One theory behind the reaction, which was popularized by the software engineer Joel Spolsky, is called the *leaky abstraction*. According to Spolsky, an abstraction (i.e., the illusion of life) is destroyed when an underlying artificial mechanism is revealed.[3] But the leaky abstraction only describes the shattering of an illusion; it doesn't explain why the revelation leads to the feelings of fear and anxiety that lie in the uncanny valley.

Numerous researchers into human-robot interaction have been studying the reactions of humans to robots in detail, including using advanced imaging techniques to observe what goes on in the mind of a person encountering the uncanny valley. One theory held by many roboticists has to do with the cognitive dissonance of encountering something that looks human but whose behavior reveals that it's not what it appears to be.

In 2011, an international team of scientists led by Ayse Pinar Saygin at the University of California in San Diego used functional MRI (fMRI) imaging to study the brains of volunteers when they watched videos comparing a human, a highly lifelike robot named Repliee Q2 (one of Dr. Ishiguro's realistic androids) and a stripped-down version of Repliee Q2 that showed its metal joints, circuit boards, and wiring.

The human and the robots all performed the same ordinary

actions: waving, nodding, picking up a piece of paper from a table, and taking a drink of water. The human's natural biological appearance coupled with natural movement were just what the subjects expected. The stripped-down version of Repliee Q2 was clearly a robot that moved with jerky, robotic movements—again, just what one would expect from a machine. The full version of Repliee Q2, however, looked highly lifelike yet moved with jerky, mechanical movements. The researchers observed their volunteer subjects' brains using fMRI while they watched the videos to see how they responded to each of these scenarios.

When the volunteers watched videos of either the human or the very mechanical-looking robot, their brain activity showed nothing remarkable. Saygin attributes this to the fact that both the human's and the mechanical robot's movements were exactly what the subjects expected them to be based on their appearance. However, when the subjects watched the humanlike Repliee Q2 perform the motions in a jerky, mechanical way, the resulting cognitive dissonance caused a little electrical storm to light up their brains. In particular, the part of their brains' visual cortexes that processes bodily movement and the part of the motor cortex that contains mirror neurons went into overdrive.

Dr. Saygin interpreted her findings as signs of the brain's struggle to reconcile a mismatch between the humanlike robot's appearance and its movement. "What [the brain] seems to be doing is looking for its expectations to be met—for appearance and motion to be congruent," she said.[4] The fact that the robot's appearance was that of a biological human, yet its movements were robotic, plunged it into the uncanny valley. The mechanical movements of the clearly nonhuman robot didn't create discomfort, because its movements were in keeping with its appearance.

Other studies have confirmed Saygin's theory that it's the disruption of expectations that triggers the uncanny effect, but why is our reaction so strong? Why is the experience of uncanniness not merely disconcerting but characterized by feelings of anxiety, dread, and even fear?

It's not only robots that have the ability to be creepy; computer-generated animation is known to have the same effect when the depiction of human characters or anthropomorphic animals and objects becomes too realistic for comfort while still failing to cross the last hurdle of perfection. Film critics have been pointing this out since photorealistic computer-generated imagery (CGI) in animation came onto the scene in the early 2000s.

The 2004 film *The Polar Express*, a Christmas tale designed to entertain children, got icy reviews from the film critic Paul Clinton. He opens his review with, "This season's biggest holiday extravaganza, 'The Polar Express,' should be subtitled, 'The Night of the Living Dead,'" he wrote for CNN.com. "Those human characters in the film come across as downright . . . well, creepy . . . So *The Polar Express* is at best disconcerting, and at worst, a wee bit horrifying."[5] The characters, meant to look hyperrealistic, fell into the uncanny valley by their ultimate lack of believability, revealed by their fixed, unfocused stares. They made reviewers think of zombies. *Newsday*'s reviewer John Anderson was even less forgiving of the character's "dead-eyed creepiness." "*The Polar Express* is a zombie train," he wrote.[6]

Other CGI- animated movies have fared no better. Even the voices and likenesses of big-name stars like Jim Carrey in the 2009 animated film *A Christmas Carol* couldn't overcome the unintended creepiness of "a dead-eyed, doll-like version of Carrey,"[7] or the Jeff Bridges character in 2010's *Tron: Legacy*, whose face was described as "an animated death mask" by a *New York Times* critic.[8]

The theme of animated corpses is a common one in people's reactions to computer-animated characters, which doesn't seem to afflict traditional hand-drawn animation like that produced by Disney and *Looney Tunes* in the past. A possible explanation for the uncanny valley is that too-realistic yet flawed animated characters and also robots remind us of death and our own mortality. Uncanny valley–like reactions have a long history and may be deeply hardwired into our brains.

The fear and dread of zombies and animated corpses has long been a common theme in horror literature, movies, and popular culture, long before the days of animation. When describing the Frankenstein creature in her 1818 book, *Frankenstein; or, The Modern Prometheus*, Mary Shelley wrote:

> It was impossible to look at it without a shudder. No mummy restored to life could be more awful than this monster. I saw the creation unfinished; it was ugly even then; but when his joints and muscles started moving, something turned out more terrible than all fictions of Dante.[9]

Implicit in the very existence of Frankenstein is the theme that a mere human body, reanimated but minus a soul, is a terrifying monstrosity. The same theme is common to virtually all popular conceptions of zombies, whose unnaturalness is seen not as a neutral feature but one of evil and destruction. Closely related is the theme of dead, soulless bodies being invaded and "possessed" by malevolent spirits who can inflict grave harm on human beings through their supernatural powers. We may be loath to admit it, but our fears of the "undead" lie deep in the recesses of our minds and are not so easily dislodged.

Mori, in clarifying his uncanny valley theory in 2012, wrote, "Imagine a craftsman being awakened suddenly in the dead of night. He searches downstairs for something among a crowd of mannequins in his workshop. If the mannequins started to move, it would be like a horror story."[10]

The assumption would be that some unseen entity had found a way to invade the mannequins to get them to carry out some malevolent acts. We might unconsciously suspect that robots can be likewise possessed. Many people have a primal fear of corpses because they project onto them frightening entities like ghosts and evil spirits. Such superstitions might have their roots in the fact that, in the past, unprotected contact with infectious corpses was known to pose a risk of contagion and death to those who handled them. However, the uncanny valley sensation goes beyond a fear of infection to a fear of death in general.

Mori writes, "Why were we equipped with this eerie sensation? Is it essential for human beings? I have not yet considered these questions deeply, but I have no doubt it is an integral part of our instinct for self-preservation."[11]

Karl F. MacDorman, a researcher at the University of Indiana, believes the uncanny experience reaches deep into human biology and psychology. "Deep in our lizard brains, where the hardwiring is kept locked up by evolution," he says, "we carry the instinct to avoid things that may cause us harm, even if that thing is projected onto a movie screen or plugged into a wall socket."

MacDorman believes that uncanny robots incite in us brief flashes of our own mortality, something psychologists refer to as "animal reminders." These flashes of awareness may not even be conscious, but they force us to confront death, reminding us that "we have a life span. We're born, we die—just like any other animals."[12] Even the most rational among us is subject to unexpected reminders of our own mortality.

Freud provided an explanation for the uncanny valley experience long before Mori described it in relation to robots, associating it with the phenomenon of animism. Animism is an ancient belief in a supernatural power that permeates the universe. It attributes the existence of a soul to practically everything—humans, animals, plants, inanimate objects, and even natural phenomena.

Animistic beliefs, according to Freud, are both a matter of cultural inheritance and a natural part of early childhood development, when babies are learning to sort through what is alive, dead, or animate. The confused emotions associated with this process become repressed memories, many of them scary or disturbing, that we carry with us all our lives.

Freud claimed, "An uncanny experience occurs either when infantile complexes that have been repressed are once more revived by some impression, or when primitive beliefs that have been surmounted seem once more to be confirmed."[13] It's easy to see how animistic beliefs lead easily to a belief in ghosts and demons and persist on a subconscious level even in the most rational adults.

"Each one of us has been through a phase of individual

development corresponding to [the] animistic stage in primitive men," writes Freud, "[and] none of us has passed through it without preserving certain residues and traces of it." He goes on to explain that among the most frightening animistic attitudes are all things "in relation to death and dead bodies, to the return of the dead, and to spirits and ghosts."[14] The widespread popularity of horror movies that shock us with grotesque images of death and the "undead," such as zombies, vampires, and, sometimes, robots, is a testament to the persistence of ancient beliefs in animism.

Bertrand Tondu, an engineer at the University of Toulouse in France, has proposed that any reminders of zombies "awake in us an animistic mind-set, which has never been abandoned. Its effect is doubly anxiety-provoking because, for a start, any resurfacing of repressed memories is, as theorized by Freud, frightening and, subsequently, this return is attached to a particularly scaring and repressed emotion."[15] A robot that appears almost, but not quite, human is also not quite alive, and this impression triggers a flash of animism along with a fleeting reminder of our own mortality.

Other theories behind the uncanny valley propose a deep, though unacknowledged, confusion about whether robots are conscious and alive to experience. Psychologists Kurt Gray and Daniel Wegner, after studying the responses of forty-three research subjects who viewed two videos, one in which a robot was extremely lifelike and one in which the robot's inner workings were clearly visible, suggested that feelings of uncanniness are driven by an attribution of mind to humanlike robots. But when we attribute mind to robots, we're still basically ignorant of what type of mind a robot might have. If we fail to sufficiently anthropomorphize them, it's easy to imagine that the robot mind is alien, inscrutable, and possibly dangerous.

Still another potential issue that a possibly conscious robot could raise is the fear of an assault on human distinctiveness. "The hallmark of humanity is our minds," write Gray and Wegner.[16] If robots can look almost indistinguishable from humans and have minds as well, we are cast into a reexamination of what, exactly, makes humans distinct, and this can create deep feelings of unease.

If we decide that robots do have minds, possibly frightening questions arise. In spite of our projections of human traits on them, we can't dispel completely the knowledge that robots are not real human beings, so what type of mind might they harbor? The idea that robot minds, even though they may have been created by humans, are still inescapably alien leads to anxiety about how to classify them and how to predict what they might do.

The question of whether robots have minds analogous to the human mind immediately makes us ask whether robots can, or ever will, be capable of emotion. The capacity to feel emotions is a fundamental human trait, one that has been lauded, celebrated, and decried by poets and scientists alike. Gray and Wegner place emotions at the very center of human experience, suggesting that "the deep-seated, implicit and intuitive essence of our minds is instead our hearts—our feelings and emotion."[17] They leave open the possibility that, whether they have real emotions or not, robots created in our image will help us understand ourselves better, writing:

> The idea of a fully human machine may be only an idea, but advancing technology suggests that there may come a time when we are swept away by deep poetry about the human condition, written not by flesh and blood, but by silicon and metal. The question is whether we will always be unnerved by the idea.[18]

Some observers of human-robot interaction, while acknowledging that engineers are focused on designing robots around the uncanny valley, have questioned whether our feelings of eeriness and revulsion toward uncanny robots will always exist or whether they're signs of a passing phase that will give way to new feelings as our familiarity with social robots grows.

The anthropologist Cheyenne Laue attempts to describe a more holistic view of robots that synthesizes the knowledge that they are both strange and familiar at the same time. She sees animism as not just a universal part of human history but a living part of the modern human psyche—witness the propensity of twenty-first-century people to attribute genders and personalities to their cars, computers, and Roomba floor-cleaning robots. "Indeed," Laue writes, "the fact that many research participants simultaneously confirm that robots are neither real nor alive but react as though they are both, indicates that a propensity for animism may be as much a part of the human evolutionary future, as it is a part of our past."[19] It may be futile to try to erase all traces of animism in human life. But maybe we can grow beyond it through everyday interactions with social robots.

What about creating intimacy in our robot relationships? Laue suggests the question of whether we will ever experience true intimacy with robot companions rests not on whether robots are conscious or are capable of feeling love for us but on the dynamics that will arise naturally from interacting with them. "We might well expect intimacy to occur simply as a natural result of growing contact and familiarity rather than as the successful engineering away of difference," she says. "In short, we may come to accept aspects of robots' strangeness as a part of whom or what these robots are, rather than perceiving dissimilarity as flaws or problems with design."

In other words, our confusion and eeriness about how to categorize social robots may be superseded by the establishment of a new category of being. Our concept of "robot" may be fraught with uncertainty now, but that may not always be the case. Laue proposes that "intimate relationships may emerge inside of the uncanny valley, inasmuch as uncanniness represents a space where familiarity and strangeness rub together, both frightening and fascinating us and pulling us forward into a future where relationships are not bounded by biology."[20]

Roboticists today can draw upon considerable research that seeks to make the human-robot interaction convincing enough to put people at ease and allow them to integrate their robot companions into their lives. There is not only an uncanny valley for a robot's appearance, there seems to be one for behavior as well.

Psychologists who study human-robot interaction have found that people of all ages are generally open to welcoming humanoid social robots into their lives, but they seem to have a "Goldilocks zone" for just how human and how perfect they want their robots to be. They want them to be neither too eerily human nor too "perfect" in their behavior, since absolute precision can be a dead giveaway that one is interacting with a machine.

A few years ago, the UK computer scientist Mriganka Biswas presented findings at the International Conference on Intelligent Robots and Systems in Hamburg that showed that people who interacted with social robots responded better to robots that were not *too* flawless in their behavior. Robots that were programmed to occasionally make judgmental errors and incorrect assumptions were better received than those whose interactions were perfect. And robots that occasionally misfired emotionally with an inappropriate response put their users at ease compared to those whose behavior was "perfect." Apparently, our notion of perfection in a robot involves a closer facsimile of ordinary human imperfection.

Some of the same qualities that put us at ease when dealing with humans also appeal to us when engineered into robots. We find the presence of certain flaws reassuring. What we also find reassuring is the presence of "personality." If we can assign personality to any inanimate object, it's the personality that draws us in, gives us a sense of familiarity, and invites us to interact. This observation hasn't escaped the engineers and programmers developing socially interactive robots. Central to creating credible personalities is the effort to endow them with emotion—or at least behavior that feigns emotions and responds appropriately to emotions expressed by humans.

Biswas and fellow researchers at the University of Lincoln studied the interactions between human subjects and ERWIN (which is the acronym for Emotional Robot with Intelligent Network), a robot with the ability to express five basic emotions, and Keepon, a small robot designed to study social development in children by interacting with them.

During half of the interactions with the robots, ERWIN

and Keepon performed flawlessly, but in the other half of interactions, the two robots were programmed with "cognitive biases" resembling those of humans. ERWIN forgot simple facts, and Keepon showed excessive happiness or sadness, which he expressed through noises and movements. The children were asked to rate their experiences, and almost all of them rated their experiences with the robots higher when they made mistakes. "The cognitive biases we introduced led to a more humanlike interaction process," Biswas said. "We monitored how the participants responded to the robots and overwhelmingly found that they paid attention for longer and actually enjoyed the fact that a robot could make common mistakes, forget facts and express more extreme emotions, just as humans can."[21]

This study suggests that people are uneasy with robots that appear *too* abled, perhaps smarter and more abled than humans. If robots' intelligence and functionality eclipse our own, that places them on more dangerous ground, where they might actually replace or potentially harm real humans. This is something researchers Francesco Ferrari, Maria Paladino, and Jolanda Jetten refer to as a threat to human distinctiveness.

The design of robot "personalities" already draws heavily on a body of research that is directed at finding ways to build empathy between humans and robots. Roboticists realize that if we're to take robots intimately into our lives, we need to be able to suspend disbelief in their realness, at least when we're interacting with them. We also need to believe, through anthropomorphism or some other mechanism, that there is some authentic empathy on both sides of the relationship. Biswas has spoken of robots using body language and a set of emotional expressions that we have been conditioned to recognize and understand since childhood. Other researchers write that our hardwired responses to social robots' expressions take their origin not just in childhood but in human evolution.

The facial expressions showing six basic emotions—anger, disgust, fear, joy, sadness, and surprise—are understood more or less universally without a language translator. There seems to be a shared inventory of easily recognized expressions that people of all nationalities and cultures understand, and roboticists are programming these expressions into social robots.

Thanks to the wealth of psychological research, the chances are that the social robots of the next few decades will be in the Goldilocks zone of robot appearance, emotional expression, and behavior, a phenomenon that will make them highly appealing. If we can get past the uncanny valley, we will be drawn into our relationships with personal robots because of our ingrained emotional natures, and resistance will be—if not futile—very difficult indeed.

3

COULD ROBOTS MAKE US MORE EMOTIONALLY INTELLIGENT?

Bert isn't an especially humanlike robot. He has long, mechanical arms with hands that can grip much like a human's, but his body resembles a stack of hardware. He has a large, square head with cameras jutting out on each side that somewhat suggest ears. It's his face, however, that your eyes gravitate to.

Projected onto a screen, the large, round LED eyes, movable eyebrows, and smiling mouth are capable of expressing a surprising range of emotions. His glowing blue eyes and ability to look crestfallen after making a mistake are especially effective at inviting you to sympathize with him, a fact that was amply demonstrated when researchers at the University of Bristol and University College London put him to work as a kitchen assistant in an experiment they named "Believing in Bert."

In the experiment, three versions of Bert were used to

"compete" for a job as a kitchen assistant, handing twenty-three participants (twelve men and eleven women ranging in age from twenty-two to seventy-two) the ingredients for an omelet. The first version, BertA, performed his role most efficiently and accurately, but showed no emotion and didn't communicate in any way. BertB dropped an egg when handing it over and attempted to rectify the situation by trying again but didn't speak. BertC also dropped an egg when handing it to a participant, but in doing so appeared crestfallen and upset. His eyebrows lowered, his mouth turned down, and he looked distressed. Since BertC could talk, he apologized for the mistake, telling the human participant how he would retry handing over the egg. At the end of the task, he asked the participant how he had done and whether or not he got the job. The researchers then interviewed each participant to determine their reactions to the various robots and to determine which one they preferred to work with.

This simple experiment yielded an important insight—that speed and efficiency, as best exemplified by BertA, were less effective than the trust and emotional empathy elicited by BertC in the assessment of the human participants.

Of the twenty-three human subjects, fifteen preferred BertC over the other versions, even though BertC took twice as long to perform his tasks than BertA. This was attributed mostly to BertC's communicative abilities and the simulated remorse he expressed over making a mistake. Participants even reported wanting to soothe BertC's "feelings" and avoid hurting them further. When BertB and BertC dropped the egg, they tried to help the robot by catching them or preventing them from falling. Their faces mirrored BertC's emotions, and they made comments such as "You couldn't help feeling sympathy when it dropped the egg. You see the face and just go 'awww!'"[1] In response to BertC asking if it got the job, one participant said, "It felt appropriate to say no, but I felt really bad saying it. It felt bad because the robot was trying to do its job." One participant even wrote "emotional blackmail" in his notes, with the word *blackmail* underlined.[2]

The researchers had hypothesized that trust and transparency were the issues that would win out when participants were asked to choose between the robots, and they were correct. BertC's apologetic words and emotional expressiveness led the participants to trust him and even to attribute to him more cognitive ability than he had. They concluded that their human participants were willing to trade a significant amount of efficiency for transparency and expressiveness, which enabled them to trust BertC more than BertA or BertB.

It was clear that BertC was able to push the participants' emotional buttons. The researchers write, "Some degree of mirroring of BertC's expression was observed in at least three of our participants which could even indicate emotional contagion, the notion that when people unconsciously mimic their companions' expressions of emotion, they come to *feel* [emphasis added] reflections of these. This is a powerful force, as emotional contagion can serve to increase understanding and provide a form of glue for personal relationships, as well as alleviating frustration and stress."[3] In other words, we seem to feel a kind of reflexive empathy for just about anything that shows even crude and simple signs of emotion. And we will forgive mistakes and malfunctions as long as we sense transparency, which leads us to trust.

The same forces that shape our relationships with people come into play more or less automatically when we interact with robots. The above experiment only underlines the fact that emotions are at the center of how we respond to the world—that we *feel* our way through life as much as, if not more than, we think our way through.

As Sherry Turkle, the MIT psychologist, has demonstrated through extensive research into children's interactions with robotic toys, it doesn't take a complex display of emotion to convince us that robots and other machines are intelligent and have feelings. Even simple expressions, joined with movement and speech, can be highly convincing. This also reflects how we deal with other humans who have very limited ability to communicate. We can empathize deeply with babies and small children when they're only able to display simple expressions and are unable to verbally express their feelings. And we can feel strong empathy for our animal companions when all they can do is bark, meow, growl, or whine. The same phenomenon applies to robots. Signs of emotional life appear to elicit empathy regardless of how simple the emotions.

There's no question that the social robots in existence today only simulate emotion. However, they do it in a way that elicits real emotions from us. Robots, whether embodied in some form or simply an algorithm-driven bot that lives as software, have passed the Turing test—convincing people in dialogues that they're communicating with humans—for years now. As for detecting emotions in us, robots are learning to recognize the full range of human emotional expressions, not just words and facial expressions but movements, posture, and gestures. They're developing forms of emotional intelligence that draw us in and invite us to interact with them as though they were living, feeling beings.

If robots are to live side by side with us in our homes, workplaces, hospitals, and the like, there are good reasons why roboticists are designing them to interact with us with emotion being not just present but part of the central interface.

Our emotional nature has deep roots in our evolution as a highly social species. It evolved alongside our socialness and is essential to living in complex societies. Emotions are the brain's shorthand for knowing how to react to a situation without much thought or analysis. They save time and energy when we need to recognize friend or foe, and they regulate social interactions. The crowning emotion for social cohesion is empathy. And empathy for robots is about to become a day-to-day part of our emotional landscape.

No doubt we will feel empathy for them, but will robots ever feel real empathy for us? Robots are being designed to behave with emotional intelligence, but will our relationships with them make us more emotionally intelligent or less so, as we depend more and more on an essentially one-sided relationship? Is it emotionally intelligent to have a companionable relationship with a robot in the first place?

In his groundbreaking 1995 book, *Emotional Intelligence*, psychologist and journalist Daniel Goleman reported that up to 80 percent of one's success in life is due to emotional intelligence—recognizing, processing, and managing emotions in one's self and in others. Goleman was inspired to write the book after reading a 1990 paper on emotional intelligence (EI) by psychologists Peter Salovey and John Mayer, who defined EI as "the use of feelings to motivate, plan and achieve in one's life." This process requires monitoring one's own and others' emotions, to "discriminate among them and to use this thinking to guide one's thinking and actions."[4]

Salovey and Mayer's article represented a turning point in a long tradition, going back at least as far as the ancient Greeks, regarding emotions as inferior to, and disruptive of, reason, and initiated a new appreciation of the positive role that emotions can play in human life. They recognized that while we tend to distrust emotions and acknowledge that they can sometimes mislead us, emotions can also tell us things that we ignore at our peril. By attuning ourselves to our emotions, we become more aware of our own and others' vulnerability. This realization, if we are cognizant of our own gut feelings, leads us to empathy for ourselves and others.

The contemporary philosopher Martha Nussbaum has identified an aspect of EI that could explain why we prefer imperfect machines over "perfect" ones—our essential vulnerability to being hurt by the external world. In her 2001 book on emotions, *Upheavals of Thought*, Nussbaum identifies the universal neediness of human beings, a neediness that arises from emotions that "reveals us as vulnerable to events that we do not control."[5] Being rejected, misunderstood, ignored, and harmed in ways big and small are inevitable in human experience, and the emotionally intelligent among us are able to make peace with their own neediness and the neediness of others. Successful relationships include an unspoken agreement to honor, and be considerate of, our mutual dependence due to our lack of self-sufficiency.

We all depend on others for their approval, affirmation, and support because all of these are needed for our flourishing. When we enter into emotional relationships with robots, whether they are embodied or exist within a computer, they might feel nothing, but our neediness comes into play. Can a machine offer us the real affirmation we need? At most, our interactions with them are rehearsals for interactions with real people.

Nussbaum's observations about our essential neediness and lack of self-sufficiency explain why we're more comfortable interacting with imperfect robots like BertC than we are with sublimely efficient, flawless machines. We find the "imperfect" robot accessible and relatable. It puts us at ease in a way that allows us to relax into a relationship that is far more similar to our relationships with real people. This phenomenon creates a seductive illusion—the illusion of a real relationship where it's an open question whether one can realistically exist.

Empathy and morality are intimately bound up with EI, and we need both to be truly emotionally intelligent. Constant striving to understand and empathize with others is the foundation for growth. Salovey and Mayer observe that "empathy is also a motivator for altruistic behavior. People who behave in an emotionally intelligent fashion should have sufficient social competence to weave a warm fabric of interpersonal relations."[6]

Robots may be able to provide some part of this warm fabric for those who lack emotionally intelligent friends and family members. But this is where robots may forever fall short of *possessing* true EI or even in convincingly feigning it. Since they don't feel emotions, they also fall short in what psychologists call "intrapersonal emotion," or the ability to recognize and manage one's own emotions. However, they could indeed enhance our own intrapersonal emotional landscape through programs that provide guided discussions aimed at helping us identify and manage our feelings.

Scientists are hard at work creating bots, whether they are embodied or are simply machine algorithms, which they hope can provide some of the same benefits as discussing one's troubles with a close friend or therapist. One example is the aptly named chatbot Woebot. Created by Stanford University AI experts with the help of psychologists, Woebot is designed to be a friend, therapist, and confidant. It operates as a subscription service and, for around forty dollars per month, checks in with you daily for a chat, tracks your moods, plays games, and curates videos for you to watch, all in the service of managing and improving mental health. It asks you questions such as "How are you feeling today?" and "What kind of mood are you in?" to prompt the kind of regular introspection that is a cornerstone of emotional intelligence.

The aim of Woebot's creators is to not only provide daily contact and maintain mental health but to actually *improve* on the work of human counselors. Woebot has a distinct advantage—you can tell it anything and the bot is incapable of judging you. Alison Darcy, one of the psychologists behind the development of Woebot and the founder and CEO of Woebot Labs, says, "There's a lot of noise in human relationships. Noise is the fear of being judged. That's what stigma really is."[7] And stigma is what we all wish to avoid when confessing our innermost secrets to another living, breathing human being, whether it's in the context of a therapeutic relationship or over a glass of wine. It's the essence of inhibition, and this fear might discourage the most disturbed among us from ever confiding, and thus confronting, their deepest concerns and issues. However, that which one is afraid to share with another person one might confess to a robot, thus bringing the issue into the light of consciousness.

The style of therapy that Woebot is designed to utilize is cognitive behavioral therapy (CBT)—a popular form of psychotherapy that challenges negative patterns of thought about oneself and the world. CBT has shown considerable success in treating conditions like anxiety and depression. The bot's approach is meant to illuminate the inner landscape so that issues can be explored and perhaps reframed in more productive ways. A daily check-in is something that a human companion may not think to do, or may do but not follow up on with further questions that challenge one's negative thinking. Woebot's creators say he doesn't come up with profound insights to tell you something you didn't know about yourself; he facilitates that process so that you can come up with your own insights—exactly what a CBT therapist hopes to do, only the very possibility of judgment is out of the equation.

From the standpoint of demand, Woebot has been incredibly successful. I spoke with Darcy about her brainchild, and she described her surprise at how quickly Woebot has gained popularity. One evening she came home from work, sat down at the kitchen table, and realized that Woebot had talked to more people in a few months than most clinicians see in an entire lifetime.

She attributes the phenomenon to the fact that Woebot has been endowed with a personality and a backstory that draws people in. "He has robot friends and a personality," she told me, "sort of like an overconcerned stepdad." And yes, Woebot simulates empathy, a skill that Darcy believes is integral to social robots. All the questions Woebot asks his "patients" are written by psychologists to help them home in on their feelings and to challenge negative, unproductive patterns of thought.

Darcy set out to design an online therapy program to fill in the gap between those who feel depressed and anxious but, for whatever reason, don't or can't go see a human therapist. "Depression is the leading cause of disability in the world," she told me, yet large swaths of the depressed population are underserved because they live in places with few to no therapists, because they can't afford therapy, or because they feel the stigma of seeing a human therapist is too daunting.

Not everyone who feels drawn to the conversational interface with Woebot is severely ill; the program is also seen as a preventative measure for those who tend toward melancholia but, with a little help, can keep themselves from plunging into the paralyzing despair of clinical depression. "We're looking toward a new model for mental health, one that seeks to maintain healthy habits that stave off anxiety and depression rather than just respond after the fact," said Darcy.

I asked Darcy what happens when someone engaged with Woebot sends signals of serious troubles, perhaps suicidality. Wouldn't a human therapist be best in such a situation, perhaps initiating medication or even hospitalization? Although she assured me there was a "safety net" in the program that would immediately refer the patient to resources that can intervene, Woebot could still fall short, in my opinion, when a real crisis is developing.

Another issue that I find troubling is that the program operates through Facebook Messenger, the messaging app hosted by Facebook. That means that the user's data is saved by Facebook (now Meta), the beleaguered organization that has previously found itself in trouble for failing to prevent the sharing of data about millions of users with Cambridge Analytica (CA). CA then mined the data to send political marketing to up to fifty million Americans. Although Facebook has faced congressional hearings and has apologized for the breach, no one to date has been totally satisfied that such an incident couldn't occur again. After all, the information users might share with an online therapist is likely to be among the most intimate and sensitive.

Dr. Darcy claims that all the information is rendered completely anonymous by the program, that Woebot seeks informed consent, and that all the data is subject to Facebook privacy terms. However, as even the most informed security experts will tell you, true online privacy has so far been an elusive goal, and I believe Facebook itself may take years to earn back the trust of its users.

While I regard the privacy issue as a stumbling block for many people before they will engage with an online therapist, I believe that programs such as Woebot can help fill in the gaps in a mental health system that fails to reach millions of troubled people, many of whom suffer in silence out of an inability to pay or the fear of stigma. In addition, even the best human therapist can't possibly check in every day with every one of his patients. There's no reason why such programs can't be incorporated into social, consumer robots, where they can help users maintain good mental health. Online bots show every sign of being, for many people, their choice for addressing issues of mental health.

There's research to back up the theory that people will share things with robots that they won't tell a human therapist. The Defense Advanced Research Projects Agency (DARPA) is deep into robotics research, and in 2014, they studied people's interactions with a virtual therapist named Ellie.

Ellie is an avatar developed by the University of Southern California Institute for Creative Technologies, and they put it to work testing 239 human subjects divided into two groups. One group was told it was talking to a bot, and the other group was told that there was a real person behind the avatar. DARPA's interest in the project was aimed at assessing the use of an avatar/bot to treat soldiers with PTSD. In the experiment, the participants who thought they were talking to a robot were "way more likely to open up and reveal their deepest, darkest secrets," according to *WIRED* reporter Megan Molteni. "Removing even the *idea* of a human in the room led to more productive sessions." Molteni goes on to say that removing the "talk" from talk therapy actually helps people to open up as well. She explains:

> Scientists who recently looked at text-chat as a supplement to videoconferencing therapy sessions observed that the texting option actually reduced interpersonal anxiety, allowing patients to more fully disclose and discuss issues shrouded in shame, guilt and embarrassment.[8]

While any progress in the evolution of psychotherapy will be good news for those struggling with mental health issues, one has to ask whether the preference for chatbots and texted therapy sessions are a sign of the type of alienation lamented by Sherry Turkle and other experts who study and write about emotional intelligence.

Might such interventions actually be harmful in the long run as our fundamental ability to relate to other people erodes over time? After all, isn't one of the ostensible benefits of psychotherapy the process of overcoming inhibitions against opening up, allowing one's vulnerability to come into the light, and growing more comfortable with our own and others' vulnerability? And are shame and embarrassment always undesirable, or do they sometimes alert us to problematic traits in our mental lives that need examination?

Suppose a person was harboring violent impulses that caused some inner conflict. He may be more amenable to confessing such thoughts to a robot, but the act of discussing this conflict with a human therapist might nudge him to be more critical of the thoughts and to work to address them. A human therapist who advocates for regular progress may be what's needed to help people persist in grappling with deeply entrenched problems without giving up when the going gets rough.

Companion robots may be programmed to address a whole range of mental illnesses and may play a part in keeping the healthy well. Engineers might program them to do things like suggesting pleasurable activities or those that lead to greater social connection as a way to regulate moods in those with mood disorders such as bipolar disorder or depression.

But can robots ever be programmed to always have an appropriate, context-related response? Take the example of the social robot Pepper, as mentioned previously, seeking to play a favorite song if he notices a sorrowful expression on his owner's face. Sorrow may be a highly appropriate emotion in some contexts. It should be recognized and examined with an eye toward how it might offer one genuine guidance in the consideration of a problem. A robot that constantly prods us to cheer up, possibly when our emotions are completely appropriate, could be seen as being intrusive, patronizing—possibly even maddening. These are just the sorts of challenges that roboticists, with the help of psychologists, are trying to overcome.

It's an open question whether robots will ever be emotionally intelligent enough to always respond to humans appropriately. The palette of emotionally intelligent reactions in human life is exceptionally broad. EI behavior extends to the full range of human expression from words and gestures to the way we dress and present ourselves. It may include the exercise of humor in one context and quiet listening in another. It entails the harnessing of our innermost feelings to help us adapt to a well-nigh limitless number of situations and experiences. It requires us to know when to quiet our emotions and when to give them expression. It entails intuition as well as critical analysis and calls for the judgment to know which to rely upon in any given circumstance. The only way robots could reach this level of sophistication would be if they felt emotions themselves. Until they themselves have emotions, they will only be able to provide us with emotionally intelligent companionship up to a point.

Critically lacking in the robot's repertoire is real empathy. Empathy depends on our ability to imagine what it feels like to be in another's shoes. Brain scans have shown that when we feel empathy, the same parts of the brain that we use to think about ourselves are activated. On a deep level, we identify with the other person, and the more we identify with her, the greater our empathy.[9]

According to University of New South Wales psychologist Skye McDonald, robots fall short of the key prerequisites for feeling empathy. She explains, "Firstly, the robot would need to have a rich knowledge of self, including personal motivations, weaknesses, strengths, history of successes and failures, and high points and low. Second, its self-identity would need to overlap with its human companion sufficiently to provide a meaningful, genuine shared base."[10]

There's another critical component of emotions that robots to date do not, and may never, share with humans: the embodied nature of our feelings. Philosophers down through the ages have debated whether one can feel any emotion without a concomitant physical reaction that affects the entire nervous system, the heart rate, muscles, and hormones.

The legendary psychologist William James argued that when we experience an emotion, "our whole cubic capacity is sensibly alive; and each morsel of it contributes its pulsations of feeling, dim or sharp, pleasant, painful, or dubious, to that sense of personality that every one of us unfailingly carries with him."[11]

Physical, visceral reactions even play a role in emotional contagion. When we witness someone crying, our own eyes grow moist, and when we hear someone laughing heartily, it's hard not to laugh ourselves. As James recognized, the mind and body form a kind of symbiosis, a feedback loop that works in both directions. Hence, if we smile, we're apt to feel more cheerful, and if we wring tears from our eyes, we will actually feel sad.

We know from research that when humans encounter a robot expressing a certain emotion, we're apt to mirror that expression and to experience emotional contagion from what we think the robot is feeling. This is another reason why we need robots that express emotions that are appropriate for the context and why we need their behavior to at least mimic emotional intelligence. It therefore matters greatly whether robots will ever experience true emotion and whether they will ever be conscious. It goes without saying that without consciousness, robots will never be able to feel anything, never mind exhibit true emotional intelligence.

There are a variety of opinions among artificial intelligence experts on whether robots will ever be conscious. One intriguing theory, called *integrated information theory*, proposes that consciousness is a by-product of structures that can both store large amounts of information and *manage* that information through connections that are dense enough to integrate the huge stores of memory.

This is something that the brain is good at but that is a major challenge for machines. Our brains are great at contextualizing everything we think about. As we integrate the enormous stores of information we have in our heads through the thinking process, we are contextualizing trillions of bits of information in ways that ultimately add up to a coherent meaning.

Neuroscientist Christof Koch believes that consciousness is intrinsic to highly complex matter and that if scientists could build a structure as complex and connected as the brain, consciousness would naturally arise. Having a software simulation of the brain (which others have hypothesized would create consciousness) wouldn't generate consciousness—you would have to actually create the structures that store, connect, and integrate information. A computer simulation simply wouldn't produce the phenomenon of consciousness. In 2014, he told *MIT Technology Review*:

> I think consciousness, like mass, is a fundamental property of the universe. The analogy, and it's a very good one, is that you can make pretty good weather predictions these days. You can predict the inside of a storm. But it's never wet inside the computer. You can simulate a black hole in a computer, but space-time will not be bent. Simulating something is not the real thing. It's the same thing with consciousness. In 100 years, you might be able to simulate consciousness on a computer. But it won't *experience* [emphasis added] anything . . . But if you were to build the computer in the appropriate way, like a neuromorphic computer, it could be conscious.[12]

In the AI community, thinking is not regarded as the same thing as consciousness, nor is intelligence the same thing as consciousness. The famous Turing test is based on Turing's question, "Can machines think?" and its answer depends on an AI being able to convince a human subject that, sight unseen, they are communicating with a person behind the technology. AI has satisfied this requirement in recent years, but no one yet claims that the same AI is conscious.

The psychiatrist and neuroscientist Giulio Tononi at the University of Wisconsin–Madison believes that, while robots could conceivably attain consciousness at some point, they will never develop real emotion. He has spent the last decade developing a mathematical framework for consciousness and believes that the brain's ability to not only integrate vast stores of information but to contextualize all that information is too monumental a requirement for a machine intelligence to ever achieve. He says, however, that emotion isn't necessary for a machine to be conscious. I'll explore this concept more in the next chapter, but for now, how might we answer the question of whether social robots will hurt or enhance our emotional intelligence through our relationships with them?

I believe that social robots that express emotions will possibly help us at the level of Emotions 101. For those who struggle with emotions, like people with autism, or those who are simply out of touch with their own feelings, day-to-day interaction with an embodied (or even non-embodied) chatbot can lead to greater awareness of one's own emotional landscape, an integral component of EI.

In addition, by modeling socially appropriate behavior, social robots can, up to a point, help us to learn socially appropriate behaviors. This might sound overly simple, but the fact is that there are many people who are at the level of Emotions 101—those who are not fortunate enough to live in an emotionally intelligent environment with family and friends who can model healthy, appropriate behavior. The high rates of mental illness, crime, spousal abuse, divorce, and all manner of dysfunctionality in our society speak eloquently to this fact.

It has been demonstrated that chatbots can indeed offer a kind of cognitive behavioral therapy for the depressed, the lonely, or those who simply have a hard time being in touch with their own emotions. There are many people for whom daily interaction with a well-designed social robot would be a step up from the dysfunctional relationships they have with the real people in their circle.

In their initial incarnations, robots will have a special impact on the lonely, children, and the elderly, who also happen to be some of society's most vulnerable. Which brings us to our next question: Just what boundaries should we impose on the use of consumer robots in the home? In other words, should you get a robot nanny? I'll explore this question in chapter 8.

4

WILL ROBOTS BE SMARTER THAN HUMANS?

Roboticist Sam Kenyon has written that we humans take a lot of our abilities for granted—so much so that we're almost completely unaware of what it takes to perform simple, everyday tasks. Take driving a car, for instance. We don't think about the complex interactions between our brains, eyes, arms, and feet because through repetition, our brains, eyes, arms, and feet just "know" what to do. But this wasn't so when we were first learning to drive. Then, every decision, every movement, had to be conscious and deliberate.

Now imagine teaching a robot how to walk, talk, and perform a variety of household tasks. Every movement involves a complex interplay of perception, decision-making, and execution. Getting a robot to perform tasks at the level of a three-year-old child would require a huge amount of programming—that is, unless the robot has the capacity to learn like a three-year-old child, by exploring the world for itself through trial and error.

BRETT, a gray-and-white humanoid robot that looks like a clunky cousin of Rosie, the housekeeping robot in *The Jetsons*, is the star of the research program at the University of California at Berkeley robotics lab. His name stands for Berkeley Robot for the Elimination of Tedious Tasks. He has a mind that is somewhere between that of an infant and that of a toddler, and he learns accordingly.

The task before him is to screw a bottle cap onto a bottle. BRETT lunges at the bottle. Even his movements resemble those of a toddler as his articulated fingers rather clumsily fail to fit the cap on top of the bottle. But each time BRETT fails, he pauses and processes exactly what went wrong, then makes another attempt. After several tries, BRETT eventually gets the cap screwed onto the bottle, something he taught himself to do with no programming whatsoever.

It's no coincidence that BRETT seems to learn just as a human child would. Programming robots to do everything that we might want them to do is a virtually insurmountable task. Pieter Abbeel, who runs the robotics group at Berkeley, was inspired by watching child psychology tapes showing how toddlers learn. Only by imbuing robots with a similar capability, he reasoned, would we overcome the problem of an almost infinite number of variables that a robot might encounter in the environment, even in a single household. Programming a robot to respond to every single challenge is impossible because roboticists can't possibly anticipate every single circumstance that a robot might encounter in the real world.

The standard for learning through exploration of the environment and trial and error is still the human brain. The brain is by far the most powerful computer, many times more powerful than the most advanced artificial intelligence, and it seems to have an almost endless ability to learn, grow, and adapt.

Endowing robots with an artificial brain that learns as it goes along has been the holy grail of robotics, and now it may be within reach thanks to artificial neural networks, a technology modeled on the human brain. The technique employed by the UC Berkeley researchers to enable BRETT to learn new tasks is called *deep learning*, and it's enabled by human-inspired neural circuitry, in which connected layers of artificial neurons process raw sensory data.

Deep learning is the technology used by speech and visual recognition systems such as Siri, the iPhone program that responds to vocal commands, and Alexa, Amazon's home assistant. Robots like BRETT incorporate deep learning as part of a feedback loop between their "senses"—cameras and other sensors—and their brains, that even includes a system of rewards when they get something right (something called *reinforcement learning*).

Deep learning is considered the most important advance in artificial intelligence since the 1950s, and household robots will be steeped in it. It means that robots can analyze huge stores of information from their environment and from the Internet, even from watching videos. The technique works so well that there are robots learning to cook from watching YouTube videos, and a robot recently managed to assemble an IKEA chair from reviewing the instructions.

The robot's thinking process is centered around layers of

interconnected artificial neural networks based on the neural networks of brains. Artificial neurons, called *units*, simulate hundreds, thousands, or millions of densely connected brain cells to learn things, recognize patterns, and make decisions. Deep-learning robots must first be encoded with key data, to have a foundation on which to analyze information by encoding the key factors of the data, translating it, in a sense, into a "language" it can understand. When the robot receives new information, it assigns a "weight" to each bit of information, and then it converts it into representations, much as we humans do. This assigning of weights leads to hierarchies of information that provide the robot's understanding of its mental representations.

The units are divided into input units and output units. When a unit receives new information, it modifies it based on the weight assigned to it and then sends the modified information to an output unit. The weight assigned to the information influences the strength of the signal sent to the next unit, much like neural networks in our brains tell synapses to develop patterns based on our thoughts and experiences. Those output units then fire and send the information to the next layer of input units, where it gets modified further, then sent to the next output unit, and so on until the information has passed through a huge number of units, modified repeatedly along the way. Following a large number of "training efforts," the robot eventually learns to make the right decision. "Eventually," however, happens incredibly fast, considering that robots can process millions of pages of information per second.

This process also provides a self-correcting feedback system, which ensures that the robot continuously refines its knowledge and never stops learning from new information, much like a human being. This process is called *backpropagation*. It means that the robot constantly compares its decision output to what was intended. When discrepancies arise, it goes back and adjusts the various weights it has assigned to bits of information and tries again. So, much like a human, it learns from its mistakes.

One of the greatest challenges for robots is identifying patterns in unstructured environments, such as a home, and in separating verbal commands from background noise in a noisy acoustical environment. This requires an ability to distinguish between signals that convey some pattern and those that are simply noise.

Another difficulty is correctly interpreting language. This leads to frustrating but sometimes hilarious results. In 2017, an African gray parrot living in London ordered a set of gift boxes from Amazon by talking to Alexa, the Amazon Echo device. The parrot's owner was mystified when she got a confirmation of the purchase on her cell phone. She was able to play a recording of the conversation between the bird and Alexa and realized that while she was at work, Barry, the parrot, had been conversing with Alexa in Afrikaans, the language he had learned to mimic in South Africa. He may have been speaking gibberish, but somehow he used the "wake word" Alexa, and the machine attempted to make sense of his sounds. Something Barry had said was interpreted as a voice command to order the boxes on Amazon. The parrot's owner said, "Barry only needs to hear something three times" to learn it, and clearly, she had used the wake-up word repeatedly.[1] Barry's little stunt proved that Alexa wakes up and listens when key words are spoken, as will all household robots, which will often be watching as well as listening.

Deep learning is not a new approach to AI; in fact, it was first proposed in 1944 by Warren McCulloch and Walter Pitts, two neuroscientists who later founded the cognitive science department at MIT. Since then, the science of deep learning has gone in and out of fashion, until the 2000s, when it exploded into what is now being hailed as a revolution.

The development that enabled this revolution—the graphics processing unit, or GPU—was first used by the gaming industry. GPUs concentrated thousands of processing units on a single computer chip, and researchers soon realized that the design is remarkably similar to the neural nets needed for deep learning. The primary innovation in recent years has been to add multiple, sometimes thousands, of layers of input and output units in deep neural nets, hence the *deep* in deep learning.[2]

AI is not exactly the same thing as deep learning, even though the terms are often used interchangeably. *AI* is a broader umbrella term that includes a number of ways to create smart machines, including machines that can be preprogrammed specifically to perform certain tasks. Woebot, the therapy-delivering chatbot profiled in the previous chapter, is an example of AI, but not of deep learning. Deep learning can be seen as a category of AI in which machines learn for themselves after being given access to data. It's the ability of robots to automatically learn from data that has led many roboticists to identify robots equipped with neural networks as the next general purpose technology (GPT)—that is, a technology with so many applications that it literally transforms our world.

There have been successive waves of GPTs (not to be confused with "generative pre-trained transformer," as in programs such as ChatGPT) over the past few centuries that radically altered science, the economy, and the social order. First came the steam engine, which led to railroads and machines of production. Then came the harnessing of electricity and its application in countless ways that improved life and led to a quantum leap in productivity. Next came the computer, then the Internet. Robots are considered the next GPT because of their ability to learn almost anything and apply their knowledge in countless ways in the real world. Deep learning multiplies their abilities exponentially.

Just consider that, like a child, a robot can learn by watching us do things. And there's evidence that robots actually "think" about what they learn from humans. Researchers at Cornell University in 2013 wrote algorithms for a robot that led to its learning several human behaviors by watching humans eating a meal. After watching the humans eat, the robot cleared the table without spilling any liquids or leftover food. After watching a human preparing to take medicine, the robot proceeded to fetch a glass of water. And amazingly, after watching a person pour milk over a bowl of cereal, the robot decided on its own to return the milk to the refrigerator.[3]

Deep-learning robots are quite systematic in the way they break down activities into several parts, such as reaching, lifting, and pouring, continuously matching new information against things they've already learned. As mentioned earlier, one of the biggest challenges for home assistant robots is being able to make sense of a cluttered environment. The more ordered the environment, the more easily the robot can function.

Does this mean that we'll have to keep extraordinarily neat homes to enable our robots to help out? For most of us, that's simply not realistic, so roboticists are focusing on enhanced 3D perception and ever-more sophisticated algorithms that will make it easier for robots to make sense out of our chaos. However, there will probably be a threshold for how messy an environment can be and still be accessible to a learning robot. Robots learn by repetition, and there must be a number of constants in their environments that they can measure against new information, such as a human's movement or an open door.

Like children, robots will learn not just from watching and emulating us but by what we tell them. We'll teach them how to do precisely what we want in precisely the desired way. Over the course of months and years, we'll have a big investment in such robots in ways that are both practical and emotional, and our attachment will deepen with each passing year. This is only one of the reasons why Yiannis Aloimonos, the director of the University of Maryland Institute for Advanced Computer Studies, says deep-learning robots "will be the next industrial revolution."[4] But robots will not only learn from us; because they're a connected technology, they'll learn from a wide variety of sources, including from each other.

The obvious example of robots learning from each other is that if a manufacturer teaches a robot to do a number of things, it can copy the robot's software into all other models. However, there's another way that your personal robot can learn from every other robot: RoboBrain.

The brainchild of Ashutosh Saxena and other researchers at Stanford University, RoboBrain is an online knowledge base that robots can both query to learn things and contribute their own knowledge to, a kind of Wikipedia for robots. Because information needs to be highly specific and broken down into simple, constituent parts for a robot to learn a new task, information in the database speaks a versatile "robot language" that can be understood by robots with all kinds of sensors and designs. It aggregates what robots like BRETT have learned through trial and error and educates robots using images, video, text, touch, and learned concepts.

Because all robots will have access to RoboBrain and can add what they've learned to it, the database will eventually be enormous.[5] So while we'll be training our robots through voice commands and example, they'll be learning continuously on their own. Over time, a robot will be able to amass a large number of skills. While RoboBrain is still being constructed, Saxena has said that in the next five to ten years, we will likely see "an explosion in the ability of robots."[6]

There is, however, what many see as a dark side of deep learning—the AI "black box" problem. What this means is that even the engineers who design the algorithms used in deep learning ultimately cannot say how, exactly, a robot arrives at a specific decision. There's a mystery at the center of deep learning that is the source of some anxiety as we anticipate a future in which robots and algorithms play an ever-increasing role in our lives and in both personal and social decision-making.

The black box problem is one reason why Carnegie Mellon University computer scientist Adrien Treuille has said, "An intriguing possibility is that we're closing the era of comprehensible science."[7] In deep learning, one ultimately inscrutable decision seems to build upon another, and engineers are not able to articulate the robot's pathway to a decision. In 2016, Jason Tanz wrote in *WIRED* that "when engineers do peer into a deep neural network, what they see is an ocean of math: a massive, multilayered set of calculus problems that—by constantly deriving the relationship between billions of data points—generate guesses about the world."[8]

Tanz writes that, in the job market, computer programming skills will no longer be so important in a world run largely by deep-learning machines. "Whatever the professional implications of this shift," he says, "the cultural consequences will be even bigger. If the rise of human-written software led to the cult of the engineer, and to the notion that human experience can ultimately be reduced to a series of comprehensible instructions, machine learning kicks the pendulum in the opposite direction. The code that runs the universe may defy human analysis."[9]

To some, the black box problem represents a dark abyss wherein control over our machines is becoming limited and indirect, a problem that will only accelerate once robots take over designing other robots and algorithms. We won't even have total control over what data is fed to our robots because of ever-changing databases like RoboBrain and the Internet. On the other hand, our relationships with our robots will become somewhat like a parental one. We'll become their cocreators, no longer needing coding or programming skills to create in them exquisite compatibility with our needs.

Deep learning has already shown us glimpses of how incredibly useful it can be, whether we can trace every twist and turn of the decision-making process or not. In 2015, researchers at Mount Sinai Hospital in New York fed the medical data of seven hundred thousand patients to a computer algorithm they dubbed Deep Patient. This data included everything from patients' test results to their medical histories. Deep Patient proved to be amazingly good at predicting who would develop a range of ailments, including cancer, and performed much better than any other predictive method. It even predicted when patients were likely to develop psychiatric disorders like schizophrenia, something that's notoriously hard for physicians to do.

The researchers who built the algorithm regard it with the kind of awe one might feel for a bright or talented child. They had a hand in creating it, but can't say exactly how it arrives at its decisions.[10] (Some decision-making algorithms can be understood based on the explicit logic and math built into the programs, but others remain inscrutable.)

Modeling deep learning on the human brain has led to humanlike conundrums like trying to explain behavior when much of the thinking process is unconscious. Algorithmic thinking might be analogous to the unconscious mind and just as ultimately murky. But this lack of transparency doesn't invalidate the usefulness of the process. So far, the trust we've placed in algorithms has been, on balance, well placed, but what do we do when something goes horribly wrong and we need to fix a grave problem? Part of the answer is to imbue machines with social and emotional intelligence to the greatest extent possible.

While building brain-like structures and functions has led to major advances, there may be inherent limitations to such an approach. Not the least of which is the assumption that, for an AI to be truly intelligent, it must duplicate the human brain and all its structures, neurons and glial cells, synapses, and so forth. One must build a perfect analog to the human brain when the brain still holds manifold mysteries that we have yet to understand. And even if we achieved this goal, it's not at all clear that a perfect analog would perform as well as or better than the brain. We may or may not create consciousness or artificial general intelligence along the lines of human ability.

One of the drawbacks of neural nets and deep-learning algorithms is that the process consumes massive amounts of energy. In fact, when in 2012 an early Google neural net taught itself to recognize cats, it took one thousand different machines and sixteen thousand processors to accomplish the task. But while research prototypes can be exorbitantly expensive, over time, the cost of computer technology generally nose-dives by the time it reaches ordinary consumers, in keeping with something called Moore's law.

In 1965, Gordon Moore, a cofounder of Intel, famously observed that the speed of computer processors doubles approximately every two years. At the same time, the size and cost of computer chips declines precipitously. Adding more microprocessors to ever-shrinking chips means that chips can make faster and faster calculations. Huge advances in the power of computer chips have taken place since 2012.

Some say that Moore's law, which held steady for fifty years, ended in 2005, but other advancements may more than make up the slack. The innovation making strides today is called *parallel computing*, and once again, the touchstone is the human brain. The brain is great at running several processes all at the same time—for instance, you can listen to a person speaking while remaining alert to all your sensory perceptions, remembering certain facts, and anticipating your response all at the same time. In AI, parallel computing means that the computational workload is shared by several microprocessors simultaneously on the same chip. The result should be that chips are getting closer to processing information in ways that mimic the human brain, and they'll need far less energy to do it. The race now is to find out how much parallelism can be loaded onto a chip.[11]

Other branches of research are focused on designing chips specifically for deep learning. Neural nets are generally written in software, which is run on traditional, energy-hungry chips. IBM recently developed a chip with the neural net written into the hardware, in actual silicon. This new chip only consumes about 1 percent of the energy that a software-based program would use, and has one hundred times the speed. This greatly eclipses the pace of Moore's law up until now and could mean a quantum leap in the efficiency and functionality of deep learning.[12]

More innovation along these lines is likely to lead to ever-faster, more powerful AI that will allow deep learning to permeate our lives, in everything from our cell phones (which already have early versions of deep learning) to our household appliances and personal robot assistants. The ability of such computing power used with neural nets is likely to be a watershed that will transform work, play, and an ever-growing array of activities that are assisted by deep learning.

With greater computing power comes more intelligence. Algorithms have been around for a long time (they were invented in the ninth century by the Persian mathematician Muhammad ibn Musa al-Khwarizmi), but it took modern computers and a quantum leap in raw computing power for AI to take off and become ubiquitous in the environment. Computer scientist Richard Mark Soley recently told *Forbes*, "Now what used to be thought of as supercomputers are inside smartphones. They cost a million times less, are a million times faster and have a million times as much memory." What used to require computer arrays the size of a city block can now be held in the palm of your hand.[13]

The evolution of robot capability will continue to follow technological improvements. Even though research into AI began in the 1950s, it languished for a few decades because computer chips simply didn't have the processing power to make it practical. At various times, AI has been widely hyped, only to fail to live up to scientists' predictions. The resulting backlash led to an "AI winter" during the 1970s and 1980s, when the subject lost respectability in academia, though subdisciplines such as AI pattern recognition and electrical engineering proceeded apace.

However, with the advent of GPUs and advanced chips, the issue of insufficient processing power is becoming less of a problem, and we're in another era of optimism about the prospects of creating true generalized AI. Experts still differ on if and when artificial general intelligence (AGI) will become a reality, but in 2012 and 2013, the philosopher Nick Bostrom polled AI experts for their predictions on when true AGI can be expected, and the average predictions by these experts were 2040 and 2050.[14]

AI has become a more respectable field in research since academic researchers started focusing on several discrete components of intelligence, such as computer vision, voice recognition, and language processing. Today, many see the potential convergence of advances in various areas, including deep learning, as possible within the foreseeable future, even if exact predictions are notoriously hard to make.

AGI is not exactly the same thing as *the singularity*, a term popularized by Ray Kurzweil in his 2005 book, *The Singularity Is*

Near: When Humans Transcend Biology. The singularity is the moment when computers and AI become as intelligent as humans, and Kurzweil predicted that this event will occur around the year 2029. Many AI researchers predict that the singularity will quickly lead to an "intelligence explosion" as AIs rapidly start to copy themselves and write even smarter algorithms. The smarter AIs then design even smarter AIs, which in turn improve upon themselves until soon they have achieved superintelligence, a level far above the intelligence of humans. It's likely at this point that no one will understand these machines. In 1965, the statistician I. J. Good famously predicted the following:

> Let an ultraintelligent machine be defined as a machine that can far surpass all the intellectual activities of any man however clever. Since the design of machines is one of these intellectual activities, an ultraintelligent machine could design even better machines; there would then unquestionably be an "intelligence explosion," and the intelligence of man would be left far behind. Thus the first ultraintelligent machine is the last invention that man need ever make.[15]

So, to answer the question at the beginning of this chapter, "Will robots be smarter than humans?" the answer isn't all that comforting. First of all, I assume that at least some household, consumer robots will be endowed with state-of-the-art AI, but it's essential to distinguish between the two major types of AI.

One is the kind of intelligence that today's machines already excel in—sifting through massive amounts of data in almost no time, recognizing patterns, and performing mind-boggling calculations with incredible speed and precision. This type of intelligence is what's referred to as *weak* or *narrow AI*, and in this domain, machines such as IBM's Watson already surpass us. If you feel threatened by a machine that can beat you at chess or calculate fractions out to the fiftieth decimal point, you might as well give up the ship; machines have already left us in the dust.

Artificial *general* intelligence resembling that of humans, though, is referred to as *strong AI*. This type of intelligence includes things like intuition, common sense, inductive and deductive reasoning, creative imagination, true emotional intelligence, and the ability to shift easily from one task to a completely different kind of task using the same information in a different context. Will machines ever dream, aspire, love, or long for things we can't imagine? While we already have some exciting research advances in instilling emotions and values into algorithms, we simply don't know. But so far, we know that no machine comes close to combining the many forms of intelligence that make us human.

It remains to be seen if superintelligent machines will ever develop the enormous versatility of humans and amount to true general intelligence. But for some, the very prospect of a machine that in any way exceeds human intelligence is automatically a doomsday scenario.

5

DO ROBOTS SPELL DOOMSDAY FOR THE HUMAN RACE?

The emergence of superintelligent machines, machines far more intelligent than we are, is rather widely, though not universally, accepted by computer scientists and technology entrepreneurs as an inevitability. Many of them regard this development as an existential risk to humankind. Digital brains could be many orders of magnitude bigger and faster than the human brain, and they could end up outperforming humans in a great many ways, including social skills.[1] The existential risk, according to those who fear AI, is that superintelligent machines will regard humans as insignificant at best and as competitors in a world of limited resources at worst. Either way, they reason, there's a reasonable chance that machines will take control of society. If enough things go wrong, they could possibly wipe out all of humanity.

Part of the visceral fear behind this doomsday prediction is the fact that machine intelligence, no matter to what extent scientists try to mimic the brain, will inevitably be alien to us. Superintelligence will almost by definition be impossible for us to understand, all the more so since machines will be creating other machines.

One risk is that an AI's goals will not align with human goals—in fact, humanity may find itself in the way of machines' goals and thereby be eradicated the way we would eradicate ants at a picnic. Or programmers might unwittingly create bugs that become amplified each time an AI designs another AI until they result in catastrophe. The danger is even greater when AIs are embodied in the form of robots such as autonomous military weapons. Other threats to human welfare have arisen recently with the release of OpenAI's ChatGPT bot, which has been shown to make what many perceive to be malicious statements and has the ability to spread misinformation at an unprecedented rate. ChatGPT's darker possibilities led more than one thousand tech experts to sign a letter in March 2023 calling for a six-month pause on the development of even more advanced AI.[2]

While many technologists wax almost messianic in their optimism about the gifts AI is likely to bestow on us, some highly influential people, including the late physicist Stephen Hawking, Tesla CEO Elon Musk, and Bill Joy, a cofounder of Sun Microsystems, have famously cautioned the world about the dangers of superintelligent AI becoming uncontrollable. In 2014, Hawking and AI experts Stuart Russell, Max Tegmark, and Frank Wilczek published an article in Britain's *Independent* newspaper pointing to an "IT arms race fuelled by unprecedented investments and building on an increasingly mature theoretical foundation." They note that

> everything that civilization has to offer is a product of human intelligence; we cannot predict what we might achieve when this intelligence is magnified by the tools that AI may provide, but the eradication of war, disease, and poverty would be high on anyone's list. Success in creating AI would be the biggest event in human history. Unfortunately, it might also be the last.[3]

They cite autonomous weapons that not only seek out targets but make decisions about when to destroy them, a technology that the United Nations and Human Rights Watch want to ban, as an example of how AI might quickly outrun human control. Referring to AI in general, they write,

> One can imagine such technology out-smarting financial markets, out-inventing human researchers, out-manipulating human leaders, and developing weapons we cannot even understand. Whereas the short-term impact of AI depends on who controls it, the long-term impact depends on whether it can be controlled at all.[4]

It's this last point that has arrested the attention of scientists and science fiction writers alike. Literature and films are littered with the idea that autonomous AI, once created, will turn on humanity, cleansing the world of a human "infestation." In fact, Hawking and Musk in 2015 joined hundreds of scientists in signing an open letter that called for a ban on autonomous weapons. One response to the letter illustrates the gaping abyss between those who, in the words of Microsoft researcher Eric Horvitz, "are providing almost religious visions . . . their ideas are resonating in some ways with the same idea of the Rapture,"[5] and those who see the possibility of human subjugation and possibly even extinction.

For their efforts, the signatories of the famous 2015 letter were given the Luddite of the Year Award in 2016 by the think tank Information Technology and Innovation Foundation (ITIF). The Luddite prize is awarded each year to highlight what the foundation considers the year's worst "anti-technology ideas and policies." When asked whether they intended to target Musk and Hawking, ITIF's president, Robert Atkinson, replied,

> Do we think either of them personally are Luddites? No, of course not. They are pioneers of science and technology. But they and others have done a disservice to the public—and have unquestionably given aid and comfort to an increasingly pervasive neo-Luddite impulse in society today—by demonizing AI in the popular imagination . . . if we want to continue increasing productivity, creating jobs and increasing wages, then we should be accelerating AI development, not raising fears about its destructive potential.[6]

So far, we're still quite a way from artificial general superintelligence. In fact, AI today is facing a paradox, says Oren Etzioni, who is CEO of the Allen Institute for Artificial Intelligence in Seattle. In 2017, he told *Popular Science* that "things that are so hard for people, like playing championship-level Go and poker have turned out to be relatively easy for machines. Yet at the same time, the things that are easiest for a person—like making sense of what they see in front of them, speaking in their mother tongue—the machines really struggle with."

Etzioni goes on to say that a major challenge is creating AI that understands not just grammar and syntax but what we actually mean when we speak. "Understanding of natural language," he says, is what sometimes is called *AI complete*, meaning if you can really do that, you can probably solve artificial intelligence.[7]

Philosophers are debating many aspects of artificial general intelligence, and one line of reasoning revolves around the Chinese room argument, a thought experiment introduced by philosopher John Searle more than forty years ago. The debate is still unresolved among those who believe that AGI, and possibly even machine consciousness, is possible and those who think AI will never truly master language comprehension, a cornerstone of general intelligence.

Searle's argument addresses two preconceived beliefs about strong and weak AI. Searle has defined strong AI as an intelligent program that transcends a mere tool of the mind and actually *is* a mind that understands the information it processes and has bona fide cognitive states. He contrasts this possibility with the belief that we're likely to be stuck with weak AI forever, a status quo where computers only emulate true intelligence by manipulating words and symbols that they don't truly understand.[8] Searle articulates his view with his Chinese room argument, which goes as follows.

Searle imagines himself as an English-only speaker locked in a room and given a large manuscript written in Chinese. However, he doesn't know a word of Chinese and couldn't even distinguish it from Japanese or any other pictographic language. He writes, "To me, Chinese writing is just so many meaningless squiggles."

Searle then proposes that he's given a second batch of Chinese writing that presents a list of rules that pertain to the first batch of "squiggles." Since the rules are written in English, they equip him to correlate the second batch of writing with the first batch.

Next, he's given a third batch of Chinese writing with rules in English that allow him to correlate the third batch with the first two batches, and this time, the rules tell him how to respond to Chinese symbols with shapes given to him in the third batch. Unbeknownst to him, the people giving him the Chinese writing refer to the first batch of symbols as a "script," the second batch a "story," and the third batch "questions." They call the rules written in English the "program." In effect, they're feeding him stories and asking him questions about the stories with instructions on how to answer the questions written in English.

Searle then proposes that, after systematically correlating these documents just by following the instructions, he gets so skilled at following the rules that anyone from the outside could easily mistake him for a native speaker of Chinese, even though he doesn't speak a word of the language.

The upshot is that the English speaker, by being given the rules, can supply the correct squiggles and create the illusion that he speaks Chinese while understanding absolutely nothing. Searle concludes that this "formal symbol manipulation" is all computers will ever do. He asserts that "the programmed computer understands what the car and the adding machine understand, namely, exactly nothing. The computer understanding is not just (like my understanding of German) partial or incomplete; it is zero."[9]

In Searle's view of digital intelligence, there's no room for volition and therefore no danger of AI "waking up" and setting out to destroy humanity. AI may be a useful tool, as Steve Jobs has said, but it will never have anything resembling a will that's capable of deciding anything. It will have goals, but those will be the goals that we give it, and we need not give it goals that would in any way, shape, or form harm humanity.

Some experts claim that, even when we give AI what we think are benign goals, it's possible that humanity may unwittingly find itself in the way of those goals, and a superintelligent AI would unerringly figure out how to eliminate any impediment to the fulfillment of its goals. A good example of the disagreement between AI optimists and AI doomsday predictors took place recently between the Harvard psychologist Steven Pinker and Elon Musk.

Pinker has claimed that the fear of unbridled AI strikes him as "the 21st-century version of the Y2K bug," which caused millions of rational people to fear that, given computers' lack of ability to assign the date 2000, at 12:00 a.m. on January 1, 2000, all the computers in the world would go haywire, causing widespread disaster. "At that moment," says Pinker, "bank balances would be wiped out, elevators would stop between floors, incubators in maternity wards would shut off, water pumps would freeze, planes would fall from the sky, nuclear power plants would melt down and intercontinental ballistic missiles would be launched from their silos."[10] In other words, AI is given the Hollywood treatment on steroids.

Pinker observes that society seems rather perversely focused on disaster scenarios that so far haven't come true. "Doomsday is hot," he says. "For decades we have been terrified by dreadful visions of civilization-ending overpopulation, resource shortages, pollution and nuclear war. But recently, the list of existential menaces has ballooned." He cites the fear of being enslaved by robots as one of several doomsday scenarios, saying, "This includes the possibility that we will be annihilated by artificial intelligence, whether as direct targets of their will to power or as collateral damage of their single-mindedly pursuing some goal we give them."

Some scientists have proposed that AI will have a will because volition will arise automatically given sufficient intelligence. But "intelligence does not necessarily translate to evil," says Pinker. "Also, if humans are able to create unbelievably smart machine intelligence, they will also be smart enough to test said technology before giving it control of the world." He then directly challenged Elon Musk, saying, "If Elon Musk was really serious about the AI threat he'd stop building all those self-driving cars."[11]

Musk fired back in a tweet saying that Pinker doesn't understand the difference between narrow AI and general intelligence. "Wow, if even Pinker doesn't understand the difference between functional/narrow AI (e.g. car) and general AI, when the latter *literally* has a million times more compute power . . . humanity is in deep trouble," he wrote.[12] Musk assumes that AGI would have consciousness and self-awareness, which would enable it to choose objectives that are different from what its creators intended. And while neither Musk nor Pinker could be described as an expert on AI, Musk cofounded an AI safety consortium and founded the organization OpenAI, a nonprofit focused on building "safe AI." In spite of Musk's general pessimism about AGI, his organization seeks to build into it certain values and restraints that would make it friendly to humans. Pinker's expertise is in cognitive psychology, which is presumably his pathway to analyzing and critiquing human belief systems.

Pinker warns against the dangers of assuming that "humanity is screwed" because such a belief could result in paralysis that would, paradoxically, prevent us from searching for solutions against the very things we fear. "If humanity is screwed," he says, "why sacrifice anything to reduce potential risks? Why forgo the convenience of fossil fuels or exhort governments to rethink their nuclear weapons policies? Eat, drink and be merry, for tomorrow we die!"[13]

One thing everyone can agree on is that the development of AGI, however near or far into the future, needs to proceed with great care. In a 2010 paper, David Chalmers, a leading philosopher of mind, laid out a possible path forward that would allow the technology to develop under certain constraints. He notes from the outset that "it is far from clear that we will be in a position to impose these constraints," but outlines them nevertheless.[14] The following are some of those suggestions.

First, we would try to limit some of the cognitive capacities of AI so that they're good at some things but not at everything. They might lack goals of their own but still carry out tasks that we ask of them. This means that AI would not be autonomous, but it would be safer, "at least if it is in the hands of a responsible controller," according to Chalmers. It's not clear, however, how long AI could remain in safe hands.

One can imagine in the case of intelligent weaponry that such arms could end up being sold on the black market to terrorists or dictators who might give the AI nefarious goals. Still, the idea of creating nonautonomous AI to take it for a test drive, so to speak, is a sound one.

Another approach suggested by Chalmers would be to imbue the AI with values coinciding with human values. He assumes that advanced AI will be "personlike to the extent that it can be described as thinking, reasoning, and making decisions." Values such as seeking scientific progress, maintaining peace, and curing disease act as constraints on human behavior, and these same values might be carried over into AIs.

Chalmers acknowledges that human values are not perfect, and the imperfections of AI creators might carry over as well. But certain overarching values, such as human survival and well-being, could be introduced into AI by direct programming or by being included in a machine's utility function. A *utility function* is simply a mathematical function that ranks alternatives according to their usefulness to an individual. As long as responsible humans control an AI's utility function, that AI should act only within its value-laden constraints. Once AIs start to design even smarter, possibly autonomous AIs, it seems at least possible that those advanced AIs may be able to alter their own values, but designing value-constrained AIs from early on seems like a good place to start.

The two above proposals make sense from the standpoint of creating AIs that are limited in their capacities and human-directed prior to making fully autonomous AI. But already the world of research, especially with the development of deep learning, is not confined to just creating such limited and constrained

AI. Robots like BRETT, though currently very limited, are designed to learn, grow, and evolve. This avenue makes sense because the effort to program a robot (or any machine AI) to do everything that we want them is not just formidable, it would result in less capable, less useful machines.

In deep-learning machines, because the education of an AI is more like the education and training of a child, robot designers will have some, but not total, control over the machine's development. Chalmers explains, "Here the final state of a system is not directly under our control, and can only be influenced by controlling the initial state, the learning algorithm or evolutionary algorithm, and the learning or evolutionary process." But given that AI will have access to the Internet, social media, and humans other than its creator, there could be unintended consequences.

The AI need not go out looking for trouble in order to be subverted. A perfect example of a computer program gone off the rails is Microsoft's Tay, an AI chatbot released via Twitter in 2016.

Tay was created to emulate the type of chat that a nineteen-year-old millennial girl might engage in on Twitter. She was unleashed on March 23, 2016, and before the day was over, she had been transformed into a Hitler-loving, foulmouthed sociopath.

Equipped with machine learning, Tay quickly lived up to her potential for learning through interaction. Unfortunately, she learned the toxic racist, sexist messages on Twitter all too well, and some Twitter users deliberately taught her to say sexually explicit obscenities. Within twenty-four hours, Microsoft was obliged to take Tay down for "adjustments." The company then deleted some of the most offensive of Tay's ninety-six thousand tweets, many of which just entailed her repeating what human users had asked her to repeat.[15] On March 30, Microsoft inadvertently rereleased Tay, who quickly started tweeting out drug-related declarations and had to be taken down again. This episode suggests that the real problem with intelligent AIs may be the behaviors they learn from humans.

When learning robots become social, they will learn not only physical tasks but a wealth of knowledge about their owners. Teaching them will to a great extent resemble the teaching of a child, and over time, if we keep the same robot for decades or longer, they will become living digital libraries of our lives. The "books" in this library will be videos, voice recordings, emails, texts, and social media postings. They will have the most detailed transcript of our lives in existence, and those memories can be saved, maybe not forever but for a time beyond our imagination.

Our investments in our robots will resemble those in our children, animal companions, colleagues, and friends. Like children, they'll learn by observing and imitating us, and we'll accumulate a *shared* history with them that will grow more valuable over time. All this means that our attachment to our robots will be not only practical but emotional as well. Robots that learn like BRETT and simulate emotions like Pepper will be more than mere appliances.

A history of our interactions with them will enable them to cater to a whole range of needs, probably more than any other human being could match. We will value them greatly for their recorded history over time, and for the unprecedented convenience of using them as an embodied general purpose technology. They will be shrines to us after we die and will be a part of a treasured family history that our progeny will be able to access for centuries to come.

It's not just important to answer the question of whether robots will end up being smarter than humans; no less important is the question of whether humans *perceive* robots to be more intelligent than themselves. Some AI experts even argue that if a machine acts as though it has general intelligence, we will consider it sentient whether it's conscious or not. If humans perceive AGI in a robot, this sets up a dynamic in which the human defers to the robot and transfers a certain kind of power to it.

While it's unlikely, in my opinion, that robots will develop a will to physically enslave us, it's likely that we'll be vulnerable to a subtler form of slavery—extreme dependence on them to do the things for us that we no longer know how to, or care to, do for ourselves. It may not be the robots we need to fear but human nature itself.

6

LONELINESS CAN KILL YOU. COULD A ROBOT SAVE YOUR LIFE?

It's ironic that in a world of over eight billion people, loneliness is epidemic. In the U.S. alone, an astonishing 47 percent of people "often feel alone, left out and lacking meaningful connection with others."[1] Twenty-seven percent of Americans rarely or never feel that they have someone in their life who truly understands them. Forty-three percent sometimes or always believe that their existing relationships are not meaningful and don't relieve their feelings of loneliness, and only 53 percent report that they have meaningful social interactions. Contrary to popular belief, young people ages eighteen through twenty-two are the loneliest of all.[2]

And loneliness can indeed kill you, just as surely as smoking or obesity.[3] It's estimated that loneliness can take fifteen years off your life and increases your risk of death by about 30 percent.

According to a recent study by Brigham Young University researchers, loneliness can increase your risk of death by up to 60 percent![4]

Far from being a benign condition, feeling alone is toxic to our physical, emotional, and even economic well-being. It has a direct impact on the body, flooding it with stress hormones and impairing the immune system. People who are lonely are more at risk of a wide range of lethal diseases, including heart disease and strokes, cancer, infections, autoimmune diseases, and even dementia. The psychological effects can be devastating, causing anxiety and depression, and experts believe it is one of the main causes of suicide.[5] Human beings are just not made to be alone. We're wired to be intimately connected with others, and it's literally a matter of survival.

Loneliness even sets us up for a downward emotional spiral that begets more loneliness. According to University of Chicago social psychologist John Cacioppo, lonely people's brains go on high alert for what they perceive to be social threats. They become hypersensitive to feelings of rejection, which they often perceive even when no rejection is intended. They tend to rate their social interactions more negatively, and this causes them to withdraw deeper into isolation. They form more negative impressions of the people they meet, reducing the chance of human connection, and their painful feelings become a self-fulfilling prophecy.[6]

Loneliness is not a matter of simply not being in the presence of other people; it has more to do with a lack of intimacy. It's well known that one can easily be "lonely in a crowd" if there's a lack of closeness and understanding. Loneliness is "an interior, subjective experience, not an external, objective condition," according to author Judith Shulevitz, writing in *The New Republic*.[7] Superficial social interaction doesn't help and it may even exacerbate feelings of isolation. You can be surrounded by people, but if there's no emotional intimacy, you can feel utterly alone. This is well illustrated by the fact that loneliness is more common in major urban centers than in small towns. Clearly, there's no shortage of people, but the modern world largely fails to create meaningful connections between them.

Although loneliness is a problem in numerous countries, the Japanese know more about it than many societies. Caught in a vicious cycle of declining birth rates, longer lives, brutally long work hours, and a decline in marriages and even relationships, there's even a word for people dying alone in their homes and not being discovered for months or years. The word is *kodokushi*, and it's something that lonely elderly people, especially, dread.

One poignant story is that of a seventy-year-old man who, while riding a Japanese bullet train, died in his seat and went unnoticed for several days despite being surrounded by people. Author Alex Hacillo observed, "Trapped in a bizarre, worldly purgatory, endlessly ferried from one bland municipal station to another, his final resting place was an untended, anonymous grave in a state cemetery."[8] In Tokyo, the largest city in the world, thousands of commuters came and went, and failed to notice the solitary man slumped in his seat.

In densely populated Tokyo, it's estimated that more than half a million young people, known as *hikikomori*, live as lonely shut-ins in their parents' houses. It's not uncommon for groups of them to meet online and to organize group suicides in the so-called Sea of Trees, a thick forest located at the foot of Mount

Fuji. Unable to find hope of ever forming relationships, their last act is a desperate outreach for some type of human sharing, even if it's a last act of despair.

The prospects for solving the social problems at the root of loneliness in Japan are not encouraging. According to a survey conducted by the National Institute of Population and Social Security Research, 70 percent of men in the eighteen- to thirty-four-year-old age group and 60 percent of women in that age group remain unattached, and of the 5,276 single responders to the survey, 30 percent of the men and 26 percent of women said they weren't even looking for a relationship.[9] Once couples do manage to come together and tie the knot, the number of children per couple is 1.39. This pretty much ensures a severe shortage of family caregivers for the elderly and the sick.

To cope with the loneliness epidemic, Japanese society has sought out some strange-sounding solutions. At least ten Japanese companies allow people to rent actors to play friends and family members, with whom to chat, go to movies and events, and to add attendance at weddings and funerals. Just having someone pretend to care for them (at a cost of twenty-eight to forty-eight dollars an hour, plus expenses) for a day gives them temporary respite from their loneliness while letting them avoid the challenges and responsibilities of having real friends and relatives.

A representative from a company called Client Partners, which rents out actors all across Japan, says its customers are "people who lack self-confidence and are particularly sensitive to other people's judgment."[10] They tend to be people who are so caught up in the downward spiral of loneliness that the effort to establish a real relationship is too emotionally fraught to even try. These people are potentially prime customers for socially interactive robots, which can provide a sense of connection without forcing them to make the effort required to truly end their isolation.

One significant Japanese phenomenon is lonely men having virtual girlfriends through a Nintendo game called *Love Plus*. The game features three virtual girlfriends a player can choose from for an emotional connection. By using a stylus, they can tap on a touch screen, and the beautiful, big-eyed girlfriends can kiss, hold hands, send flirtatious text messages, and even get mad if they feel ignored. Unlike real girlfriends, they're at the beck and call of players twenty-four hours a day. The typical user of the popular game is a single man in his thirties or forties who has given up on finding real love.

Some of these "relationships" have lasted for years and, contrary to expectation, the choice of a girlfriend is not necessarily based on looks. Loulou d'Aki, a Swedish photographer who has profiled various users of *Love Plus*, asked players what they were looking for in a woman. "I thought they would tell me all these physical things, like 'she has to look like this,' but nobody said anything like that," she says. "They wanted someone who accepted them as they were."[11]

One of the obvious appeals of the game is that rejection is impossible and the game-addicted player will never be obliged to leave his risk-free emotional comfort zone. More recently, game developers have started creating similar games aimed at lonely women, and it will be interesting to see how many women will be drawn into these ersatz relationships.

In 2016, Vinclu, a Tokyo-based tech company, took the virtual assistant a leap forward and has turned it into a romantic partner represented by a hologram of a girl floating in a glass tube. The character is called Azumi Hikari, and she can not only turn the lights on and off but also send affectionate text messages throughout the day, making users feel connected and "loved." At a price of $2,700, the comely assistant is now sold out in Japan and the U.S.[12]

Perhaps one of the most unusual phenomena even for Japan is the powerful devotion that some men feel for Hatsune Miku, the virtual anime character created in 2007 by Crypton Future Media. The virtual character, a large-eyed sixteen-year-old girl with long blue ponytails, has been a cultural phenomenon in Japan for more than a decade. A singing, dancing animation of her appears projected on giant screens at concerts, and she presides over a $100 million industry of branded merchandise and products with her image. For $2,800, one can buy a desktop device that has a holographic image of Miku suspended inside a glass dome. It's this talking, dancing hologram that a thirty-five-year-old Japanese man married in 2018.

Forty guests attended the wedding between Miku and Akihiko Kondo, a Tokyo resident who confessed he had been in love with the character for ten years. Kondo told *The Standard* newspaper that not all his relatives attended the unofficial ceremony, but that "I've always been in love with Miku-san. I never cheated on her, I've been thinking about her every day."[13] At the wedding, the "bride" was represented by a stuffed doll in Miku's image. Now he sleeps beside the doll each night, a wedding ring worn on her left wrist.

Miku's hologram converses with him about the small intimacies of a real relationship. She wakes him up each morning, welcomes him home from work each evening, and tells him when it's time to go to bed at night. She also does all the things that an ordinary AI assistant would do, turning the lights on in his apartment after he alerts her that he's on his way home.

With his virtual relationship, Kondo doesn't feel that he's alone, and he says that after some bad experiences, he wouldn't even consider a relationship with a flesh-and-blood woman. His dependence on the holographic Miku means that he will likely not develop the emotional skills necessary for a real relationship, even if he wanted one. So far, he's satisfied with the harmonious, unchallenging connection with a simulated character that will never argue, make demands, cheat, or reject him.

Kondo sees his "marriage" to Miku as an example of a form of sexuality that should be accepted by society as just another, valid choice. On the subject of anyone trying to talk him out of his attachment, he says, "It's simply not right, it's as if you were trying to talk a gay man into dating a woman, or a lesbian into a relationship with a man . . . I believe we must consider all kinds of happiness."[14]

Kondo's story might be unusual, but it's not unheard of. Gatebox, the company that makes the desktop hologram of Miku, has sold 3,700 marriage certificates for what it calls "cross-dimensional" relationships. I can't help but think there is something sad about a human being limiting himself to a virtual relationship. The preoccupation is likely to exclude the kind of real love, with its highs and lows, its joys and sorrows, that make us grow beyond the immature beings we start out as. While Kondo grows old, gets sick, experiences the death of family members, and has to meet the normal vicissitudes of life as a mortal being, will the perpetually adolescent Miku continue to meet his needs? Will Kondo ever be able to attain real emotional maturity? And, never venturing into a relationship with a real woman, will he even know the difference?

It should come as no surprise that Japan is on the leading edge of using robot companions to battle loneliness and isolation. In chapter 7, I will delve more deeply into romantic relationships with embodied robot partners, which promises to offer much more than the captive image of a hologram, and the effect this could have on marriage and society. But for now, I'll focus on the many shades of companionable love, as they are unfolding in Japan and several other technologically advanced countries.

One of the first groups to enjoy interacting with robots is the elderly, and the government of Japan is actively promoting eldercare robots as a partial solution to its caregiver shortage. Because of the rapid aging of its population, the nation will need an additional 380,000 senior care workers by 2025, a need that is not currently being filled through traditional channels. It's estimated that by that year, there will be seven million people suffering from dementia and needing intensive, specialized care.[15] Of special concern for those with dementia is the possibility of social isolation, which can lead to rapid deterioration. But unlike the perpetual feedback loop of holographic relationships, eldercare robots are being designed to actually promote the social interaction that slows the advance of dementia.

Shin-tomi Nursing Home, located in the heart of Tokyo, utilizes twenty robots of differing sizes, appearances, and abilities to provide various services for its residents. They resemble furry animals, children, adult humanoids, and some, like the Tree, are simply utilitarian, guiding and coaching those who have trouble walking.

The robot Pepper, who was introduced in chapter 1, leads Shin-tomi residents in singing, conversations, and exercise. Some of the robots, such as Aibo, the robotic dog, and PARO, the furry baby seal, fill a role similar to that of an emotional support animal. When petted, PARO turns its head, blinks, purrs, and emits recorded sounds of a real Canadian harp seal. PARO is one of the most popular robots at nursing homes. When interviewed, one Shin-tomi resident said, "When I first petted it, it moved in such a cute way. It really seemed like it was alive," she said with a laugh. "Once I touched it I couldn't let go."[16]

One of the more visually disarming robots at Shin-tomi is the Telenoid, which looks weird, if not downright scary, at first glance. Its bald head and expressionless face are a bit eerie, and it looks as though its arms and legs have been amputated.

Only when its function is revealed does the design make sense, for Telenoid is made to sit in the lap and to promote interaction among dementia patients. It's controlled remotely by a tablet in the hands of friends or family members who want to communicate with the patient.

When the remote tablet user speaks, the Telenoid speaks, and its head and mouth move. Sometimes the tablet is controlled by nursing home staff, who use it to stimulate conversation and engagement. The bland exterior of the Telenoid allows the patient to imagine what various speakers look like, and one resident at Shin-tomi said that patients using the robot were more inclined to speak freely and to share their feelings than they were with human caregivers.[17] Even though one is interacting with a real person, the robot's appearance helps to suggest that one is actually interacting with a robot, along with the liberating sense that one cannot be judged by some of the more intimate things said.

A number of studies analyzing the use of social robots in nursing homes suggest that the robots actually lead to greater social engagement among users. One Shin-tomi resident, following an exercise session with Pepper, said, "These robots are wonderful. More people live alone these days, and a robot can be a conversation partner for them. It will make life more fun."[18] In addition to providing a conversation partner, social robots do seem to prompt more interaction among nursing home residents by not only loosening up inhibitions but by giving them something to talk about.

One of the drawbacks to eldercare robots is the cost. The robots employed by nursing homes can run anywhere from $3,800 to $8,000 each. One of the most expensive ones isn't a social robot at all but a robotic bed made by Panasonic that splits into two, one side turning into a wheelchair. The robot can assist paralytics and people with mobility problems. There are other robots to assist the elderly and disabled that can help them walk, lift them from the bed, help them take medicines, bathe them, and assist them in multiple ways.

So far, about five thousand Japanese nursing homes currently employ eldercare robots, but the trend that is so appealing in Japan may take longer to take off in the U.S. or some other societies. The reason is differing attitudes toward robots, and one of the major drivers of attitudes is culture.

One survey conducted in 2018 found that over 80 percent of Japanese people have a positive view of robots being used for eldercare. A bit surprisingly, less than half of respondents said they would rather be cared for by another person. This reflects the ambivalent feelings that most of us have about being cared for by another person. People often report that they feel that being taken care of is a burden on the caregiver, and most of us feel uncomfortable having someone assist us with intimate activities like bathing. By far, most people place a high value on independence and their ability to take care of themselves, and this is no different for Japanese society.

The Japanese newspaper *Nippon* reported that 51.3 percent of responders to the above survey said that "not having to worry about a robot's welfare was a plus," and 27.2 percent said they felt "more reticent" around human caregivers even if they provided better care than a robot.[19]

Robots can already offer the elderly a broad array of services, including mobility assistance, remote monitoring, communication with others, bathing assistance, housecleaning, fetching things, leading them in exercises, controlling other electronics and appliances in the home, vital sign monitoring and calling emergency services when needed, checking the weather, reminding them when it's time to take medicines, and playing games. All these things can help an elderly person live independently in their own home, but the emotional component is just as crucial. Some people don't have the emotional capacity to live alone. A robot that can supply companionship and even psychotherapy in addition to all these practical benefits could be an invaluable addition to an older person's home. They would also engender a level of dependence that could far outweigh expectations.

It's not only the user of the robot that could become dependent; family members and friends could become overly dependent on having a robot care for their loved one and become less engaged. Furthermore, the robot's personality is designed to please, while human relationships are inherently challenging.

It's possible that the user could come to prefer the pleasing robot over his family and friends, withdrawing more from human relationships and in reality becoming more isolated. Like the lonely men who pour their attentions into a hologram, the user's real social skills might atrophy.

One reason the Japanese are so accepting of robots is because in Japanese culture, robots have historically been portrayed as being cute, lovable, and harmless. Japanese movies, comics, video games, and books are replete with examples of cute, helpful robots, while American culture is teeming with *Terminator*-style robots and androids that inevitably rise up and want to kill their creators. In recent decades, Japanese culture has become saturated with a phenomenon called *kawaii*, which loosely translates into "all things cute and childlike," and this has had an effect on the kinds of robots being designed.

Started among Japanese schoolgirls in the 1970s as a form of rebellion against the onerous expectations of straightlaced adulthood, *kawaii* was embodied in their style of dress. They drowned themselves in ribbons, ruffles, bows, toys, and small stuffed animals—a style characterized by cuteness overload tinged with the macabre or, perhaps, a hyper-expressive Disney character on LSD. Soon cars were decked out in similar fashion and the aesthetic of cuteness spread to everything from household appliances and online games to government health pamphlets and dish sponges. As the ethnographer Sophie Knight has observed, "Japan now even prints cartoon bunnies on reminders for cancer screenings, tsunami warnings and insurance brochures."[20]

There's a science behind why we respond to cute objects, aside from the pleasant escapism from the grimmer aspects of life.

The mere sight of cute, round things that remind us of babies causes our brains to release a flood of "happy hormones"—more specifically, the love hormone oxytocin. Like many street fashions before it, kawaii bubbled up to the mainstream and has become a large part of what Japan considers a major export and a source of soft power in the world. Anime, characters like Hatsune Miku, Hello Kitty clocks, and *Pokémon GO* are prime examples of kawaii exports.

Not to be outdone in the universe of cute artifacts are the many childlike robots that have sprung up in Japan. Interacting with them creates something similar to the kitten-fueled oxytocin high of watching cat videos. PARO is one example. Another example is RoBoHoN, the robot phone. It not only walks and talks, responding to what you say, it has a projector on its head for showing pictures and films. Only in Japan, perhaps, could a robot phone whose head you hold to your ears while you speak into its feet become a popular adult consumer item. In fact, the phone is so popular that RoBoHoN devotees regularly meet up at a café in Tokyo to share their mutual enthusiasm for the robot.[21]

Another one of Japan's cute robots is Toyota's Kirobo Mini, which reacts to users' emotions and engages in conversations. The small android, with its round head, enormous eyes, and oversize feet, which speaks in Japanese, was the first companion robot in space. It blasted off from earth in a rocket on August 4, 2013, spent eighteen months at the International Space Station, then returned to earth aboard a cargo supply spacecraft, splashing down in the Pacific Ocean on February 10, 2015. Its mission was to relieve the loneliness and boredom of astronauts during a long space deployment.[22]

Kosuke Tatsumi, who is a product designer at the Japanese firm Yukai Engineering, says that making cute, childlike robots means that they are seen as approachable and that they can easily be forgiven if they make a mistake. If Sherry Turkle is right in her assertion that with robots, nurturance is the killer app, some of these cute Japanese robots will end up in households, tending and entertaining the young, caring for the sick and the aged, and comforting the lonely. As these robots invite a nurturing response, they engage the emotions and increase the attachment of the user.

So what about the argument that robot relationships will be harmful because they entail only one-sided emotion, no matter how adept designers get at making models that feign feelings? Some scholars have pointed out that the relationship is in essence artificial and even propose that it's inherently unethical. Since the introduction of social robots is too new to have amassed a large body of research into their long-term effects on users, a few scholars have examined the issue from the standpoint of the effects of fictional books and movies. After all, we all become emotionally engaged with fictional characters who can't return our feelings because they don't exist. Is this somehow bad for us?

The Danish philosopher Raffaele Rodogno has intricately explored this argument in a 2015 paper published in the journal *Ethics and Information Technology* titled, "Social Robots, Fiction, and Sentimentality." He notes that to date there is evidence that interacting with robots "increases positive mood, diminishes loneliness, alleviates stress, increases immune system response, and even decreases existing dementia."[23]

The word *sentimentality* is defined as a tendency to willfully distort reality in order to indulge certain, possibly false, and unjustified feelings. Some thinkers, most notably Robert Sparrow, find sentimentality to be deplorable because we have a duty to ourselves to "apprehend the world accurately."[24] He further asserts that the design and manufacture of robots such as the robot dog Aibo is unethical because it encourages us to indulge in false emotion.

Rodogno defines sentimentality by quoting the British philosopher Mary Midgley: "Being sentimental is misrepresenting the world in order to indulge our feelings." It's asserted that by indulging our feelings, we must betray a deeper need—the need for seeing the world as it is.

There's no doubt that when interacting with social robots, so far, at least, the robot feels no emotion. Interacting with it involves some level of make-believe in which we act as though the robot has an interior emotional life. We know that it doesn't, yet we act as though it does, willingly suspending our knowledge that the robot is a mere plastic-and-metal construction. Some level of pretense is needed for us to actually be engaged in the process. But do we betray ourselves when we imagine that the friendly robot really likes us or that the robot dog really needs our attention?

Rodogno argues that there's no reason why we have to lie to ourselves or misrepresent reality in order to derive real benefits from robot relationships. To explore this question of self-deception and real versus unwarranted emotions, he considers the process of being emotionally engaged in fictional books and films. People regularly have genuine emotional reactions to the travails of fictional characters that don't, strictly speaking, exist outside of an imaginary world. We knowingly engage in the suspension of disbelief when following the ups and downs of fictional characters like Anna Karenina or Oliver Twist. While some philosophers regard these feelings as "quasi-emotions," Rodogno explains that our feelings of sadness for Anna, for instance, are genuine emotions because people like Anna exist in the world, and our sadness easily transfers to them.

While we know that Anna is a fictional character, her pain is actually experienced by countless real people for whom we can feel compassion. And the best fiction, written by talented authors who have deep insight into human life and behavior, can teach us about much that is real about the human condition. There's real emotion involved in the realization that Anna's foibles and vulnerabilities are much like our own and that we're not insulated from tragedy. Knowing that we too are capable of making serious mistakes from time to time arouses compassion in us for those who bring trouble onto themselves. These emotions, when awakened, can even contribute to our emotional growth.

Rodogno acknowledges that in robot relationships, as in caring about fictional characters, we knowingly engage in a bit of make-believe, but that our emotions are authentic and reflect realities in the world that we can transfer to real people and animals.

He does make a distinction, however, in the case of the more vulnerable among us, including those with Alzheimer's disease or other cognitive impairments. These individuals may not be knowingly engaging in a bit of pretense; they may be genuinely fooled into thinking that the robot is alive and has real emotions. I would include on that list small children and some on the autism spectrum, whose concerns I'll address in an upcoming chapter.

Even the very lonely and those lacking in social skills may, while ostensibly knowing that the robot is not alive per se, be drawn so deeply into the relationship that it edges out any possibility of forming relationships with real people or pets. Over time, and considering that social robots will remember everything we tell them and will respond accordingly, the robot relationship may come to be seen by the vulnerable person as equally real as their engagement with other people. But in some ways, the competition is not a fair one. Real people are not always physically or emotionally available to us, and the perpetual availability of the robot could tip the scales in its favor as a primary relationship. Couple that with the fact that robots will never judge or reject us, and the attraction is even stronger.

In most circumstances, Rodogno ultimately compares emotional engagement with a robot pet like Aibo to being emotionally engaged with a good book or movie. He says, "Just as my sadness for Anna Karenina involves my imagining, accepting, mentally representing or entertaining the thought (without believing) that certain unfortunate events have occurred to her, my joy at the robot pet involves my imagining, accepting, mentally representing or entertaining the thought (without believing) that it is happy to see me." The thought is conscious, and there is no real misrepresenting of reality because even as we entertain the possibility that the dog loves us, we remain ultimately grounded in our knowledge that the robot dog isn't real. So in Rodogno's view, there is no corrupting sentimentality, because there is no false emotion. However, there's still some question about the sum of benefits from regular engagement in a robot relationship. Rodogno says of the person involved:

> The situation of these individuals would parallel that of those who spent much of their lives immersed in their novels or watching their favorite TV series. While literally speaking not misrepresenting reality, some would say that these individuals are in some sense living a life at some remove from reality. And while not misrepresenting reality they may engage in these activities in order to indulge in some feelings (or perhaps to avoid other feelings that they would other-wise have to confront).

There are other reasons why too much interaction with a social robot could be bad for us. We are all to a great extent a social and emotional work in progress, with a competing mix of healthy and unhealthy tendencies. Nothing about interaction with the robot would spur the user to examine or adjust social and emotional maladjustments, as human relationships—especially long-term relationships—tend to do. This may be an appealing feature of robot relationships in the beginning, but hardly leads to the kind of personal growth that comes from the serious self-reflection involved in negotiating the vagaries of human relationships. We have to accept that robot relationships have their limits and that making them central to our lives will likely stunt our growth and provide us with little impetus to truly flourish.

Rodogno still isn't fully persuaded that those in ersatz relationships aren't being sentimental (in a bad way) to the extent that they willingly censor some aspects of reality in pursuit of certain emotions. One is still subordinating the truth to the need to evoke desired emotions, such as affection, acceptance, and caring, without working hard for them. But even so, we have to recognize that the robot's user needed to feel something, whether it's acceptance, humor, or the sense of being needed. By using a robot to evoke these feelings, can we really say that the interaction is hurting the person? If a lonely person finds temporary relief in chatting with a robot, couldn't that relief have some salutary effect on their physical and emotional health?

Some might argue that we frequently use real people and animals to comfort us, to ease our loneliness, and to evoke positive feelings. But people and animals command real respect in their own right and will sooner or later assert their own needs. A robot will not, because it has no real needs beyond a working power supply. It doesn't need to be respected because it's not alive and has no feelings to be wounded or slighted when we treat it with insensitivity or abuse.

If we become too accustomed to using robots to elicit good feelings in us while being asked for nothing in return, we could unconsciously transfer the same expectations to other people, failing to properly respect them, and feel confusion and frustration when they rebel against our demands. This question of whether, and to what extent, habits developed during long-term engagement with a robot will transfer to real people and pets has not been adequately studied and is still an open question.

It could also be argued that a person could become so used to the habit of self-deception and replacing reality with a species of make-believe (i.e., that a robot loves us unconditionally) that he fails to respect his own emotional well-being and his need to live an authentic life. It might even be considered a kind of self-harm or self-neglect when one's real psychological issues are conveniently avoided and allowed to fester. None of this allows a lonely person to make true progress in the need to end real isolation. The person will have failed to flourish in the arena of social relations, one of the most important aspects of life.

There's another issue at stake here, and that is the staggering number of real people who are lonely, uncared for, and in need of nurturing connection. When one's attentions are being poured into a robot, they are not being poured into others who might actually need them. For Rodogno, robot relationships become unethical when they displace real people or pets. Furthermore, he notes that it's through our relationships that we find much of our identity and our sense of meaning in life. While a robot may elicit superficially pleasant feelings, it's unlikely that it would provide a deeper sense of meaning or become a large part of our identity.

In the conclusion to his paper, Rodogno asserts that "ersatz companionship may indeed make us emotionally childish in some sense, incapable or at any rate unable or unwilling of loving other human beings with all their 'imperfections,' of negotiating those difficulties that often make a bond between friends or lovers even stronger or deeper." And he notes that the issues outlined here could take on an entirely new meaning if the phenomenon of robot relationships becomes widespread, while there's way too little long-term research to see what we as a society are getting into.

It can be accurately said that most of the social robots available to date don't offer anything close to the emotional complexity of a good book or movie, never mind a deep relationship with another person. Most of them keep their conversation responsive but superficial. This isn't true of chatbots like the Woebot or some other, more ambitious conversational programs that can be incorporated into robots. But because robots are taking off as a consumer item, they will need to get up to speed quickly to meet the needs of the marketplace in the event that people demand more from them.

They can easily house programs that offer psychological therapy and other services that engage people on a deeper level than that of a simple robot dog that asks to be petted. This will especially be true of robots designed to become romantic partners. AI-infused robots will take certain forms of intelligence to a new level, one that can be intellectually challenging and demanding for those who want it. But they will not demand good treatment, complain if misused, or even be capable of rejecting us if we behave badly or fail in our side of the relationship.

If social robots really do become as common as televisions or cars, for instance, it would entail unprecedented social change. If they become the training ground for real relationships, the intimate human relations that form the bedrock of society could shift into something of a different character. Millions of people could be stuck in a kind of perpetual social immaturity, unwilling or unable to emerge as emotionally mature adults. They may be so hooked on the superficially pleasant feelings of interacting with robots that they're unaware of the deep emptiness they have inside due to their incapacity for real human connection. And they may treat others as they treat their robots, a scenario in which no one really gains anything of value.

7

LOVE IN THE TIME OF ROBOTS

Created by engineer Douglas Hines for his New Jersey–based company TrueCompanion, Roxxxy the love doll prototype was five feet, seven inches tall, weighed 120 pounds, and, her maker claimed, could express love and carry on a conversation. More than just a glorified sex doll, the prototype for Roxxxy was so realistic that she had a heartbeat and felt warm to the touch. Her speech-recognition software could "understand" human speech and respond with a number of preprogrammed responses. Potential owners could customize Roxxxy's appearance and even choose one of five personalities, from "Frigid Farrah" to "Wild Wendy."

Interacting with Roxxxy was not without its distractions. Her motors whirred continuously, and her facial expression was frozen in what can only be described as an unsettling vacancy, like an undead Stepford wife. There was no danger of an uncanny valley for Roxxxy; rather, she brought to mind necrophilia and women objectified to a disturbing extent. Roxxxy was debuted in 2010 and was still on the market in 2017 but, as of this writing, her fate is unknown; the TrueCompanion website that featured her has been taken down, and a search was unable to turn up any new information. However, newer sex robots are rapidly being developed by companies like RealDoll and Sex Doll Genie.[1]

Robots that offer not just sexual love but all kinds of relationships are in development around the world. Because social robots are being introduced as a consumer item, marketers are finding lucrative niches—sex and romance—and making the robots as seductive as they can possibly be given the current limitations of technology. The creators of love robots know that we are not exactly blank slates that they need to appeal to. We all want to be loved, admired, and flattered. A large body of research confirms that we want to be seduced and that we're vulnerable to the kind of endless, worshipful attention that robots can provide but that few humans can.

Current robots for sex and romance leave a lot to be desired in terms of realism—they would never be mistaken for real people—but far more realistic models are expected to become available in a few years. These robots will invite even deeper attachments because of their more sophisticated ability to simulate the behavior of a human lover. For those who have problems with real relationships, these robots may be the path of least resistance. But some researchers think they could provide a superficial sense of connection, only to cause more social isolation in the long run.

Key to these robots will be their ability to mimic the nuanced emotional aspects of a love relationship while demanding nothing in return, an equation that no human lover can deliver with the kind of unwavering reliability of an algorithm.

Stories of men falling in love with artificial women have become familiar fare to science fiction fans through TV shows like *Westworld* and movies like the 2013 *Her*, in which the male protagonist falls in love with an interactive computer program voiced by the actress Scarlett Johansson. In 2021, the excellent German film titled *I'm Your Man* and featuring the superb British actor Dan Stevens presents a nuanced tale of a romance between a highly skeptical archeologist and a love robot that she's been called upon to evaluate for the university where she teaches. This film focuses almost entirely on the companionate aspects of the relationship to show that, quite apart from sex, relationships with highly realistic robots can provide a measure of happiness for the lonely.

Alma, the archeologist who only reluctantly agreed to try out Tom, the robot played by Stevens, is convinced going into the experiment that she could never become attached to a robot and regards the whole enterprise as a waste of time, if not utterly ridiculous. She's hyperaware of the artificiality of Tom's constant attentions and sees the whole relationship as merely a communication feedback loop with herself. She tells a colleague that the robot is "just an extension of me." This is never really refuted, but over time is counterbalanced by the endless efforts by Tom to aid, assist, and show affection.

Tom is intelligent, sensitive, and intuitive in his efforts to please Alma. Intellectually, he is her equal, only with the ability to read and digest massive amounts of information in a millisecond. He uses this ability to help her in her research career at the university. But Alma is persistently unreceptive to his attentions and argues with him that he's just a machine, and therefore, they couldn't possibly have a real relationship. She tells him, "There's an insurmountable gulf between us," and "I'm acting in a play, with no audience. I'm acting for myself, even knowingly talking to myself." However, over time, Tom's steady, patient efforts to draw close to her wear away at her skepticism like water slowly carving its way through a rock.

In the end, Alma and Tom develop an intricate relationship, and Tom becomes everything Alma ever wanted in a man. She has to admit that the relationship with him is in many ways satisfying and that she would have a hard time returning him to the university, where the plan is to erase his memory of the time he spent with her. She suddenly finds this untenable. She starts to recognize that although the experience was not with a real person, it was still of great value.

At one point, Alma and Tom encounter on the street Alma's colleague at the university, a shy older man who is having a trial run with a female robot, and she's standing by his side. The fembot is indistinguishable from a real woman, and the two of them act for all the world like happy lovebirds. Alma and her colleague exchange a few words about how their trial runs are going, and he says, "I never thought I could be this happy. They [the robots] make us happy, and what could be wrong with being happy?" The intellectually gifted Alma is stumped.

Nothing in human experience can match the attraction, frustration, fascination, confusion, agony, and ecstasy of a romantic relationship. At the same time, relationships with real partners, even the painful ones, can spur deep personal growth and make us into better, more empathetic people. The exertions of achieving and maintaining intimacy with another person force us to get outside of ourselves and to consider the needs of others. Some would say that our naturally slothful species, averse to strenuous effort, would seldom grow into true adulthood unless spurred by necessity. But given the endless complexities of romantic relationships, would it just be easier to outsource this role to a machine that will never judge you, cheat on you, lose interest in you, or be dissatisfied with you?

The kind of sex robot that could "love" a human being would have to be very highly advanced compared to any robot existing today. But the fundamental technology, the very matrix of making this equation work, has already been developed. It's in the algorithms that are programmed to ceaselessly please the owner and to learn over time every nuance of the human partner's needs and desires. This is the crux of the seduction. Nothing the human partner ever does is objected to because the robot can't complain. The human's needs are forever center stage and are the only thing that matters. But can a robot, even a very highly advanced one, truly love its owner?

The question is one of the current conundrums being discussed by philosophers. There's no doubt that current AI-enabled robots can *simulate* some aspects of love, but the possibility of a robot ever *feeling* love for its human partner rests on whether an AI will ever attain anything like consciousness, an unresolved question. However, it does seem possible that a robot could offer a sophisticated simulation of love, and for some people, this might be enough.

Fantasy is always a strong component at the start of any romantic relationship. In human relationships, as time goes by, the feeling shifts to a more knowledgeable footing through experience with the lover. But robot relationships are bound to remain stuck in this early period, while the human partner erects an elaborate fantasy about his or her robot lover. For deep attachments to form, the human partner must project a personality, a backstory, and supposed needs and desires on the robot. Those fantasies, of course, will meet the needs of the human, who is in a feedback loop with himself. But the more specificity he can bring to his fantasy, the realer the relationship will seem.

Being in a relationship with any artificial agent, be it a hologram, a doll, or a robot, necessitates the suspension of disbelief the same way we suspend disbelief when we enter a movie theater and for the next two hours embrace a fictional story as if it were real. It requires suppression of the knowledge that the artificial partner isn't real and in fact can feel nothing toward us.

For a robot relationship to work, we need to believe that the relationship is not only real but special, particular to only this robot and us, that it has unique value. The human partner must ignore that the robot is one of a series of more or less identical machines that, if interacting with a different owner, would exhibit the same exact level of interest, companionship, and feigned love. Its algorithms would work identically no matter who owns it. Compatibility, one of the hallmarks of a successful human relationship, is not a thing to be cherished or celebrated—the illusion of it is created by impersonal algorithms.

It's the fantasies that the human partner brings to the relationship, and the feelings the person invests in the robot that can take on a life of their own. The robot's alleged feelings are not real, but the human's feelings are, and this is what has led many observers to recognize that it's impossible to simply dismiss these bonds as mere delusions. But problems remain that make the incongruity with reality concerning in the long run.

Love robots are likely to appeal to people who are already ready to invest their emotions in partners like love dolls and holograms, but for others, the need for mutual love complicates the picture considerably. As enshrined in literature, movies, and popular songs since the beginning of recorded history, a highly valued aspect of love relationships is the sense of a special compatibility between two partners who are "made for each other." An ever-accommodating love robot may over time come to seem as though its actions are based on the worthiness of the human partner, but such is not the case.

In a recent paper made available online by MIT Press Scholarship, Sven Nyholm and Lily Frank suggest a way around the "made for each other" problem by suggesting that "a robot might not be custom-made to love you in particular, but might instead be endowed with a general capacity to 'fall in love' should the right human come along."[2] But the authors admit that the robot would have to be extremely sophisticated, even to the point of self-awareness. Until that happens, it seems likely that, in the minds of most people, a robot relationship could be seen as less valuable than a relationship with a human partner because there is none of the freely occurring compatibility that can seem so magical between two people.

The philosopher of technology Mark Coeckelbergh, who has written extensively about robot relationships, places a strong emphasis on the freedom of a human partner to accept or reject us. He notes that a deep need in any relationship is the need to feel that one has been chosen above all others when rejection is a real possibility. To be chosen is to feel deeply validated, and it lends true uniqueness to the relationship. But a robot partner can't actually choose us any more than they can reject us. They are simply programmed to please us in every way. The fact that the robot has no choice but to behave in a certain way devalues its supposed devotion and actually reinforces the idea that in reality, we are not special. And since the robot can't reject us, we have no need to embark on personal growth as a way of maintaining the robot's affection. Furthermore, we are not motivated to exercise empathy in the relationship, something that human relationships hinge on.

One aspect of the robot love relationship that gives pause to experts is its asymmetrical nature. The human partner has all the power and need not consider the needs of another, and the human partner could start to see this as the only acceptable equation in a sexual relationship. Nyholm and Frank say that one party, the robot, is made for the other and does all the accommodating. The human partner is the only one who is free to be him- or herself, to make demands, or to exercise choice and dictate the terms of the relationship, a situation with an unsettling resemblance to slavery. The human partner can become so habituated to this type of relationship that he or she transfers the same expectations to human partners, setting the stage for real-world dysfunction and abuse.

Feminists in particular find this equation disturbing in the way it echoes the old-fashioned asymmetrical relationship between men and women. One could argue that women could turn the tables by having their own submissive love robots, but in practice, women have shown far less interest in having a robot partner than men have. The substitute of a robot for a person doesn't seem to play into female sexuality or the emotional natures of women, but then again, almost all the research and development has been into fembots that exist to satisfy men. It's always possible that this could change, but so far, the development of utterly submissive love robots has made them a more natural fit for men who are unwilling or unable to embark on a relationship with a woman who has her own needs, desires, and free will. Feminist thinkers are asking, "Is this really something that we want to reinforce?"

In the search for an understanding of future robot relationships, long-established features of the sex trade have folded easily into the transactional nature of a man purchasing a robot for his gratification, and this has led many to compare the love robot phenomenon to prostitution. Women throughout history have been treated as commodities through prostitution, marriage practices, and even the spoils of war. The ability of women to make choices in their love lives is a relatively recent development, and there are still parts of the world where they are frequently married off without their consent.

The ability to purchase and use fembots will do nothing to challenge the sexual exploitation of women, and some say it actually exacerbates the problem by further normalizing it. Some observers point to this as a reason that men having relationships with robots is bad for women, children, and society in general.

Always at the forefront of consideration is the effect of the robot on a human partner over time, especially if that effect spills over into human relationships. What is likely is that the robot will become more like a human, and the human will become more like a robot—unfeeling, with a limited range of social expression and emotional intelligence.

The human partner is likely to get used to the limited possibilities of a robot lover that, for all its efforts to accommodate, will never surprise or truly delight him. He might get stuck in the emotional rut of a low-risk relationship, lulled into complacency by the knowledge that the robot can never reject him. The one-sided relationship could easily become the "new normal" in one's list of preferences, refocusing one's expectations in a way that makes emotionally risky human relationships seem less attractive. A human partner's unpredictability and demands could seem burdensome and not worth the trouble.

Sherry Turkle says, "Relationships with robots are ramping up; relationships with people are ramping down."[3] In her groundbreaking 2011 book, *Alone Together: Why We Expect More from Technology and Less from Each Other*, Turkle reveals her findings from years of research into young people's interactions with robotic toys and with technology in general. She has written about a widespread trend among young people to use technology as a shield between themselves and others, retreating behind it rather than dealing with the risks and demands of unmediated human interaction.

The more one retreats behind technology, the more one's social skills deteriorate, Turkle says, and actual human interactions become even scarier. Over time, real human engagement carries such an emotional risk—the risk of failure or rejection—that it occasions major anxiety. We start avoiding human contact because it just seems easier to settle for superficial connection. Thus, the preference for social media over in-person meetups, for texting over a phone call, and using an emoji over an attempt to describe one's feelings in words. But this is a self-perpetuating cycle that feeds on itself; real social skills atrophy from lack of use, insecurity balloons, and soon the prospect of real human interaction can seem way too threatening to take a risk. Loneliness and alienation grow.

Another issue to consider is voluntary commitment to the relationship. As Nyholm and Frank point out, commitment, something we prize from human partners, is a moot point with robots. If the robot is made for you or developed over time using algorithms that cater to your every whim, it would be ludicrous to think of the robot as having willingly committed itself to you. It simply has no choice because it's incapable of choosing.

The free will of the romantic partner, his or her capacity to refuse to commit, makes the giving of a commitment a precious and ideal part of love that must be given anew countless times over the course of a relationship. To remain committed is to continuously invest in a relationship while the possibility of not doing so is always present. This is part of the agonizing yet exhilarating uncertainty that lies at the heart of every relationship where freely choosing partners continuously affirm their devotion.

The committed lover offers love and support not just when it's convenient or easy, or when it satisfies their own needs of the moment, but over time and under any kind of circumstance. It's one of the most treasured aspects of a love relationship and is one that a robot, short of developing self-awareness and free will, cannot give. While a robot may always "be there" for you, its reliability will never satisfy the deep yearning for a partner who has chosen to commit to you over every other and who willingly renews that commitment on a daily basis. You are simply not special to the robot. As Nyholm and Frank say, "A robot programmed to stick to you like a fly on a piece of sticky tape is not a lover but something else."[4]

Another complexity of the love relationship is the degree

to which it intersects with our deepest feelings of self-esteem. Whether we admit it or not, we rely on our lovers to a great extent to affirm us and to prove to us that we're lovable. In this sense, relying on the affirmations of a robot that can do nothing other than love us can't touch the deepest part of us that routinely questions whether we are lovable or not. To pretend that the robot's acceptance is enough is a species of self-deception akin to using a drug to make us feel temporarily good while we're truly miserable inside. If we're truly miserable inside, the robot may have a superficial palliative effect, but its affirmations will not move the needle.

Of course, there are people who will content themselves with the attentions of the robot out of habitual avoidance of their own emotions or because of the risks entailed with human relationships. No doubt there are those who find a kind of comfort zone in emotional isolation, and there are those whose proclivities and behaviors are likely to alienate human partners or prevent connection in the first place. For them, the utility of obtaining impersonal sex minus all the trappings of a bona fide relationship could be enough. After all, there are people who use other people this way and are completely okay with it. Some say that this type of person is better off callously using a robot rather than callously using other people.

David Levy, the British computer specialist who made a splash with his 2007 book, *Love and Sex with Robots: The Evolution of Human-Robot Relationships*, thinks that sex robots will be an acceptable alternative for men who are too shy or, for other reasons, are unable to form connections with real women. In an interview, he told *Scientific American* that love robots could be the answer "to all those who feel lost and hopeless without relationships, to let them know there will come a time when they can form relationships with robots."[5]

To put it into these terms makes the love robot seem like a compassionate solution to a social problem, yet for all the issues discussed above, would the real solution not be to address why these individuals have trouble making connections in the first place? Perhaps the problems encountered by the lovelorn can be remedied, or perhaps there really is a potential human partner out there willing to overcome the problems. After all, the Germans have a saying: "For every pot, there's a lid." This may or may not be true, but without a doubt, the person who decides to simply settle for a relationship with a machine is less likely to make further efforts to resolve their difficulties in finding a human mate.

I don't by any means intend to sound flip or to diminish the difficulties some men have in connecting with others but to explore whether these difficulties could be surmountable through therapy, emotional growth, better social circumstances, and life experience. I've lived long enough to know that many problems that positively bedevil us in our younger years can be overcome with time and sustained effort. But we often need to be challenged to leave our ruts and to put in the effort that leads to personal growth. If we foreclose any challenge to change, it makes it far harder to develop the kind of skills needed to solve a multitude of problems.

Levy has frequently compared the use of sex robots to the use of prostitutes. He says, "You know the prostitute doesn't love you and care for you, is only interested in the size of your wallet. So I think robots can simulate love, but even if they can't, so what? People pay prostitutes millions and millions for regular services." He's right, of course, that prostitution is prevalent around the world and perhaps always has been, but it would be naive to ignore the real problems of paid sex work. A major one is that prostitutes are frequently trafficked, beaten, and abused, and not infrequently murdered by men whose sexual impulses are bound up with violence.

This is not intended to be a thorough examination of sex work in general, but the comparison of sex robots to prostitutes bears some analysis. My first point is that prostitution is not necessarily the "victimless crime" Levy seems to suggest. It's no secret that some men use prostitutes because it's the only way they can vent certain desires that their wives and girlfriends will not submit to, and girls, boys, and women often enter into prostitution because of desperate life circumstances that leave them little or no choice.

Levy spends considerable time in his book discussing the prostitution angle vis-à-vis robots, and I agree with him that the use of sex robots could spare real sex workers some of the more degrading and dangerous aspects of the world's oldest profession. But I think he far overstates the prevalence of prostitution in an effort to suggest that strictly utilitarian sex is very commonplace.

Levy notes that Alfred Kinsey's 1948 report, *Sexual Behavior in the Human Male*, showed that approximately 69 percent of white American males had used a prostitute at some time in their lives but that those who used prostitution as their *only* sexual outlet was only between 3.5 and 4 percent.[6] However, Levy notes that more recent studies suggest that a far smaller percentage of men—perhaps 18–20 percent—have used a prostitute in their lifetime. A 2000 study found that only 8.8 percent of men reported they had paid for sex at some time during their lives.

Studies from around the world vary, but they commonly show that the percentage of men whose *only* sexual outlet is paid sex is quite small. The vast majority of men who have visited sex workers still have sex within the context of relationships, whereas men who adopt dolls as a sexual outlet are far more likely to forgo relationships with women altogether. Levy's implication that men who are accustomed to paying for sex will have no problem embracing sex robots is probably correct, but the number of men who don't pay for sex far outnumbers those who do.

If the number of women paying for sex is any indication of whether they will embrace sex robots, it's not likely male sexbots will be all that prevalent. Levy attempts to address this issue through the lens of prostitution, but he overemphasizes the phenomenon of women paying male prostitutes for sex. He devotes two and a half pages to men paying for sex and seven pages to the vanishingly rare occurrence of women using prostitutes, a lopsided ratio that suggests that women are far more open to utilitarian sex than previously thought. At least, this was my impression, so I did a bit of research to find out if women paying for sex is a common thing, or a thing at all. What I found provided some validation to my initial hunch that it's a rare phenomenon.

A survey conducted by researchers at University College London between 2010 and 2012 found that only 0.1 percent of British women reported that they had paid for sex, while 10 percent of men said that they had at some time done so. The highest number of women using male prostitutes that I found occurred in a 2010 study conducted by IBISWorld, a research company working in Australia. The study found that 6 percent of Australian women reported they had paid for sex at some time.[7] But that number is so much higher than the numbers cited in other studies that one can only speculate that the phenomenon of women paying for sex is either more prevalent in Australia than in the rest of the world or the study is somehow flawed.

Levy's emphasis on the tiny number of women paying for sex is representative of a common thread in discussions of sex robots, where it is often posited that male sex robots for women will be just as prevalent as fembots for men, but so far there's not much evidence to support that idea. If, in fact, Levy's suggestion that women's use of sex robots will be analogous to the use of prostitutes is accurate, then statistics regarding prostitution point to quite the opposite.

It has been suggested that men whose proclivities include the need to degrade and abuse female partners may vent these urges on robots rather than real sex workers. This is certainly possible among men who feel shame about the practices and would be more inclined to indulge their more questionable tastes with robots. But feminist thinkers caution that men who get used to inflicting abusive behavior on fembots could transfer those behaviors to real women who happen to enter their orbit. They would become so habituated to abusing the robot, which is incapable of complaining, that they would expect their human partners to accept abuse without complaint as well.

An important 2015 paper by the psychoanalyst Danielle Knafo, "Guys and Dolls: Relational Life in the Technological

Era," deconstructs the psychology of love relationships with technology. The abstract introducing the paper says it all: "This paper asserts the current age as perverse, a social paradigm facilitated by explosive technological progression that is rapidly altering the erotic and social dimensions of relationships."[8] Knafo believes that the unhealthy psychological issues of doll attachment carry over into attachments to other technological products, a phenomenon that is sometimes called *technosexuality*.

One thing seems certain, and that is that while relationships with love robots may resemble things that have long existed to some extent, they will create entirely new types of relationship by virtue of being far more sophisticated than any love doll or sexual aid known today. And by their interactive nature, they will invite complicated feelings and elaborate fantasies that will take on a life of their own.

Comparisons of men's relationships to dolls and holograms have their limits when we're talking about highly realistic love robots. The more humanlike these robots become, the deeper they will draw in their users, and the effect will be cumulative over time. I can foresee a time when the conversation shifts beyond whether such attachments are advisable or healthy to whether these robots should have rights similar to those of a real person.

The reason I think this is because the robot's rights will be seen through the lens of a relationship that involves real feelings on the part of the user. The robot may not be real, but the feelings invested in it certainly are. It's by virtue of the human emotional investment that it may one day become illegal to "kill" a sex robot by turning it off or erasing its memory. And given how devoted some technosexuals are to their artificial love objects, this group may become more active in advocating for robot rights. I'll explore the issue of robots and whether they should be considered persons in the last chapter, but suffice it to say that love robots are likely to be some of the first artificial agents to be considered for personhood.

Playing into what I see as an inevitability is the fact that modern society is much better at deciding *how* to advance technological advancements than it is at sorting out whether it *should* pursue new technologies. A huge issue here is the effect of love robots on a marriage when one or both partners outsource sex and emotion to them. Would a married person's use of a sex robot constitute adultery? Much, very much, depends on whether that spouse of a person involved with a robot feels betrayed and whether the attention lavished on the robot takes something away from the marriage relationship. But it stands to reason that attention given to the robot is attention not given to the marriage partner.

Knafo believes that the humanization of machines goes hand in hand with dehumanizing people. She bases this on studies of the attitudes of men who are attached to high-end love dolls. It's a common theme among these men to dehumanize women while creating extravagant fantasies anthropomorphizing their dolls. In her view, the emotional rut that people get into when their affections are focused on a doll is hardly a benign condition—it leads to greater alienation than what they experienced before they started to lavish their affections on a nonliving object. They may at the deepest level be uncomfortable about their choice and seek to justify it by blaming real women who, in their view, drove them to such an extreme.

Knafo is concerned about our culture's "progressive eradication of the boundaries between plastic and flesh, wire and artery, computer and brain. This explosive time heralds a new kind of life, where machines become more like humans and humans more like machines."[9] Relationships with artificial agents by their very nature blur the boundaries between fantasy and reality. This is already true of love dolls, but the robots in development will represent fantasies on steroids. They will be far more seductive than an inanimate doll.

It's conceivable that in the distant future, humanoid robots will in at least some ways be indistinguishable from humans. If or when that happens, the boundaries between fantasy and reality will be even more blurred, but the fundamental issues will remain the same. People will have deep, complicated attachments to their robots. No one really knows what effect this will have on marriages and other relationships. Will some people become so habituated to the lopsided attachments to their robots that human relationships become too demanding, too unpredictable, and too much emotional risk? Will people still be willing to do the heavy lifting that any true relationship entails?

Douglas Hines, the founder of TrueCompanion and the inventor of Roxxxy, the interactive high-end sex doll, claims that his love dolls will have a beneficial effect on human relationships and society at large. He told CNBC in 2015 that "Roxxxy provides physical and sexual pleasure but also provides interaction and engagement. It's customizing technology to provide a perfect partner—she's not meant to replace a real partner but is meant as a supplement."[10] What Hines seems to gloss over is that fact that no one lives in a vacuum but in an intricate web of relationships with other people. If the robot is a "perfect partner," human partners could pale in comparison.

Another issue is that the habitual use of a robot with no needs and desires means that the user is accustomed to totally self-centered sex and becomes less accustomed to considering the needs of others. This is another reason why one might choose sex with a robot rather than their mate. Some experts have proposed that robots could be used as sex therapists for people who have sexual difficulties. While I can see a role for robots in this regard, the robot therapy would only be truly successful if one transitions pretty quickly to relations with another human being. It's hard to see how a robot therapist could improve sex beyond the most rudimentary level, because one is not dealing with another being with needs.

While people undoubtedly project human traits onto nonhuman things, they also find it easy to project nonhuman, machinelike traits onto other humans. When one's concept of humanity is to attribute machinelike traits to them, there's no need for empathy or consideration of their rights and feelings. Studies of men who frequent prostitutes have shown that often, in their minds, the prostitute is not a person but a thing. There is no consideration of her inner experience; she is simply seen as an object.

This kind of objectification of females isn't limited to sex work. It permeates our culture and allows men (and even some women) to avoid having empathy with women and girls in general. It is the bedrock of sexism, misogyny, exploitation, and abuse. It doesn't follow that all men who grow used to using fembots for gratification will make strong distinctions between robots and the real women they encounter.

Feminists point out that fembots that have no needs and no interior life reinforce sexist attitudes and play into culturally conditioned sexuality for many males. Certainly not all men treat women as things, and our whole society is evolving in ways that enable us to see the full-fledged humanity of women. But we still have a long way to go before women cease to be objects of exploitation. And it's not likely that sexbots will do anything to break the chain of dehumanizing behavior toward women, because they will be steadily reinforcing the idea that a dehumanized partner is not only acceptable but perhaps even preferable.

Then there's the issue of personal and emotional growth. At the very least, robots will fail to spur the kind of personal growth that makes life truly satisfying. Much of the issue revolves around the emotional intelligence and insight of the human partner using a robot. If the user regards the robot as a sex toy and expects nothing more than sexual release with them, and doesn't carry this attitude over to human partners, there may not be any harm.

However, research has affirmed over and over again that humans are powerfully driven to anthropomorphize robots and easily get sucked into what they imagine are complex relationships with them. But since the robot can't make any demands or have a say in how it is treated, there's a danger that this will become a new normal that ends up leaving the users more isolated, alienated, and emotionally stunted than before. It is also likely to perpetuate the objectification of women, because men are so accustomed to dehumanizing sex with a robot.

While the rest of society is actively moving toward more equitable gender roles, the sex doll industry is certainly not dissuading anyone from seeing women as objects. But this doesn't have to be the case. In the words of Kathleen Richardson, the British ethicist, co-opting technology in a way that furthers exploitation is not inevitable. We are the creators of the robots, she has said, and we have choices about how we design and utilize them—even how we choose to classify them.

Sex robots are in the early stages of development, and we need far more research into how they will affect individuals, families, and society before they become commonplace. Otherwise, market forces will determine what role they play in society, and market forces are notoriously indifferent to the greater good. "It's important that we have a debate about this," says Richardson. "Einstein's theory of relativity—it didn't have to turn into the atom bomb."[11]

The introduction of sophisticated love robots will be a huge social experiment that interjects advanced technology into the nucleus of human relations, the romantic relationship that is so central to the life of the average man and woman. To some people, a robot will never be more than a paltry substitute for a living, self-aware, freely choosing partner. But to others, these robots will literally redefine the romantic relationship into something other than what it has always been: an adventure in emotional risk and uncertainty, the source of indescribable happiness and shattering heartbreak.

8

IS THERE A ROBOT NANNY IN YOUR CHILDREN'S FUTURE?

Would you leave your small child in the care of a robot for several hours a day? It may sound ludicrous at first, but think carefully.

Robots that can care for children could be a godsend for many parents, especially the financially strapped. In the U.S., 62 percent of women who gave birth in 2016 worked outside the home, and day care costs are often exorbitant. In Washington, DC, the most expensive place in the country, the annual cost for day care for a single child averages over $24,000. In most states, the annual cost of childcare is more than college tuition. The price varies from state to state, but in all states, it accounts for a hefty chunk of the typical family's budget.[1]

"We're talking about the Holy Grail of parenting," says Zoltan Istvan, a technology consultant and futurist. "Imagine a robot that could assume 70 percent to 80 percent of the caregiver's role for your child. Given the huge amounts of money we pay for childcare, that's a very attractive proposition."[2]

Both China and Japan are on the leading edge of employing specially designed social robots for the care of children. Due to long work schedules, shifting demographics, and China's long-term (but now defunct) one-child policy, both countries have a severe shortage of family caregivers.

Enter the iPal, a child-size humanoid robot with a round head, huge eyes, and articulated fingers, which can keep children engaged and entertained for hours on end. According to its manufacturer, AvatarMind Robot Technology, iPal has been selling like hotcakes in Asia and is now available in the U.S. The standard version of iPal sells for $2,499, and it's not the only robot claimed to be suitable for childcare. Other robots being fine-tuned are SoftBank's humanoid models Pepper and NAO, which are also considered to be child-friendly social robots.

iPal talks, dances, plays games, reads stories, and connects to social media and the Internet. According to AvatarMind, over time, iPal learns your child's likes and dislikes and can independently learn more about subjects your child is interested in to boost learning. In addition, it will wake your child up in the morning and tell him when it's time to get dressed, brush his teeth, or wash his hands. If your child is a diabetic, it will remind her when it's time to check her blood sugar. But iPal isn't just a fancy appliance that mechanically performs these functions; it does so with "personality."

The robot has an "emotion management system" that detects your child's emotions and mirrors them (unless your child is sad, and then it tries to cheer him up). But it's not exactly like iPal has the kind of emotion chip long sought by *Star Trek*'s android Data. What it does is emotional simulation, what some would call *emotional dishonesty*, considering that it doesn't actually feel anything. But as noted previously, research has shown that the lack of authenticity doesn't really matter when it comes to our response to what appears to be emotion. Children, and even adults, tend to respond to "emotional" robots as though they're alive and sentient even when we've seen all the wires and circuit boards that underlie their wizardry.

The question is whether the relationships we develop with robots causes social maladaptation, especially among the most vulnerable—young children just learning how to connect and interact with others. Could a robot in fact come close to providing the authentic conversational back-and-forth that helps children develop empathy, reciprocity, and self-esteem? Also, could steady engagement with a robot nanny diminish precious time needed for real family bonding?

It depends on whom you ask.

Because iPal is voice-activated, it allows children to learn by interacting in a way that's freer and more natural than interacting with traditional toys, says Dr. Daniel Xiong, cofounder and chief technology officer at AvatarMind. "iPal is like a 'real' family member with you whenever you need it," he says.[3]

Xiong doesn't put a time limit on how long a child should interact with iPal on a daily basis. He sees the relationship between the child and the robot as advantageous in many ways, though he admits that the technology needs to advance substantially before iPal could fully take the place of a human babysitter. iPal won't feed, bathe, or change diapers for your child, but there are other robots in development that are being designed to do these tasks, says Xiong.[4] iPal is designed for social interaction and teaching, especially in the area of STEM subjects—science, technology, engineering, and math. Robots like iPal are meant for the classroom as well, and iPal even has a tailored autism package that can be used in special education or at home with a child on the autism spectrum.

Some of these robots, like the child-size iPal and KASPAR, are humanoid in form, while others are made in the form of animals like the popular PARO, the baby seal robot being used in some childcare settings and in nursing homes for older people. All of them are cute, appealing, and fun for children to interact with.

One of the most important features of social robots, according to their makers, is their ability to consistently model the types of behaviors a child needs to learn. This can help any child learn to recognize things like social context, body language, and emotional expression.

Many studies have suggested that children learn better with robots because of heightened engagement. Voice-activated robots are simply fun and entertaining to interact with and to learn from. In an ideal world, they would help to lead a child to ever-increasing levels of knowledge and maturity. A good example of such a program is the software that has been developed to help robots provide education and therapy to children with autism.

Children born with autism spectrum disorder (ASD) can fall into a broad range of symptoms and disabilities. Some are profoundly impaired while others can be high functioning; consequently, there's no one-size-fits-all therapy for them. Common effects of ASD include trouble reading social and emotional cues, repetitive behaviors, and a restricted range of skills, often with a laser-like focus on specific interests. They can even be savants in areas like music, art, or engineering while having difficulty with many of life's simple tasks of day-to-day functioning.

There's no cure for ASD, but there are successful therapies that can dramatically improve the way people with this condition function. Teachers and therapists must tailor their efforts for each child based upon his or her deficits and abilities. A robot that treated each child on the spectrum the same would hardly be effective, but social robots can be programmed, through software and real-life interaction, to provide highly customized therapy.

An iPal that comes downloaded with the autism package and meant for use in educational settings runs in the neighborhood of $8,000, a significant cost for schools and families who want to use it at home. But surveys of autism experts have suggested that these specialized robots can have a beneficial effect on children in the areas of social and interpersonal interaction and play, and even in ways that promote their emotional well-being. One such study that examined the effect of the social robot KASPAR, programmed to work with autistic children, had beneficial effects in a broad range of areas.[5]

On its website, the Luxembourg-based emotional-robot maker LuxAI highlights aspects of robot therapy that particularly fit with the way autistic children engage. Interacting with humans is fraught with anxiety for the autistic child, and LuxAI claims that by eliminating the social anxiety that gets in the way of learning, the child is freer to focus on goals like developing basic social skills as a kind of bridge to more appropriate human interactions.

Robots especially appeal to autistic children because (a)

they're fun and entertaining, and (b) they provide predictable, consistent responses over and over without losing patience or judging the child. This way, the child is allowed to benefit from repetition until the skill is mastered, however long it takes. The robot consistently models social skills like making eye contact, understanding facial expressions, and turn-taking, all while initiating fun games and other forms of learning.[6]

In 2018, a coalition of researchers from five academic institutions released results of the first study that tested robots used with autistic children over a period of thirty days in the home setting. Similar studies had been done, but only in laboratory settings, not in the children's homes, where they could interact more naturally.

The children were engaged with the robots for thirty minutes a day through storytelling and games, giving the robots a framework within which to model appropriate social and emotional skills through back-and-forth engagement. One of the most useful features of the robots' programming was the ability to continuously adapt to the child's way of learning. In games, the program was progressive; once a child made it to a certain level, a new challenge was introduced. This worked to keep the children engaged and learning.

Brian Scassellati, the lead researcher of the study, said, "These are kids who have years of experience with the idea that social interaction is challenging and something they don't understand. When they interact with the robot, though, it triggers social responses but it doesn't trigger a lot of the other baggage they've come to associate with social interaction."[7]

The children's social and emotional skills were tested both before and after the thirty days, and, according to Scassellati, the children in his study "showed improved performance across the board." The children's caregivers (all either parents or grandparents) told researchers that, among other things, the children were better at making eye contact and were more able to initiate communication. One worry on the part of the researchers—that the children would become bored with the robots and disengage before the thirty days were up—didn't happen. The children not only stayed engaged, but the improvements they had made were still maintained when tested thirty days after the robots were removed from the home.

While studies of children and robots differ in their interpretation of how helpful social robots are with teaching healthy social behavior, they agree that children are highly engaged when interacting with a robot and that they find the encounter enjoyable and entertaining, two factors that enhance learning.

One advantage to home teaching / therapy robots is that it would be impossible (and unaffordable) for any human caregiver or teacher to be as consistently available as a robot can. There's no limit to how long a robot can interact in a day's time. Autonomous, embodied robots will be limited only by their battery time. For autistic children, it seems reasonable to assume that specially programmed robots could at the very least enhance and accelerate progress initiated by human therapists in helping the child master basic social and emotional expression.

Nevertheless, the accomplishments attained with the help of the robot should be transferred to human interaction to be truly successful. It's the very seductiveness of social robots that troubles many people, who think that human relationships should be the gold standard for developing children. After all, it's in the arena of human social interaction that the child must be able to function. The trick is to use technology as a side dish, not an entrée. The robot interaction, it's hoped, can be a piece of the childcare puzzle without replacing the human teachers and caregivers who form the bedrock of a child's care and education.

The limited emotional range of current social robots could be a valuable and effective step in getting the child from one level of development to another, but they should at some point be outgrown by the child. The level of expression in robots with hard plastic faces can model rudimentary emotions, but those skills should be transferred to people (and pets) once they are mastered. At the moment, the robots providing autism therapies lack the expressive complexity needed for a child to progress beyond a certain point.

That will almost certainly not be the case as the technology of embodied robots evolves. There are ongoing efforts to develop hyperrealistic robots with a large range of facial expressions, but this area is still plagued by the challenge of overcoming the uncanny valley. One example is the robot Ameca, the eerily expressive robot made by the UK company Engineered Arts. Some people love Ameca, but many (like me) find her unnaturally animated, over-the-top facial expressions to be the stuff of haunting nightmares.

It seems that teaching and therapy robots for those on the autism spectrum have some true advantages as an add-on to the labors of human teachers and therapists. But what about teaching the more normative children and young people using embodied, AI-enabled robots in the classroom?

One reason for developing teaching robots is that it can both augment learning and address a shortfall of teachers. There is a shortage of teachers worldwide today, and there are several factors driving this development. One of the most-cited ones is the decline in the numbers of young people electing to go into education.

This is not just an American problem. School systems everywhere are struggling to recruit teachers.[8] Bosede Edwards and Adrian Cheok report in a 2018 paper published in *Applied Artificial Intelligence* that the numbers of students enrolled in programs preparing them for a career in teaching has dwindled dramatically in recent years. The low-pay, low-recognition reality of teaching jobs are impediments to recruiting young people into the field and are contributing to "educational deserts" in rural and underserved areas, and in urban centers, remote villages, and territories in developing countries.

Could the use of robots take some of the workload off teachers, add engagement among students, and ultimately invigorate learning by taking it to a new level that is more consonant with the everyday experiences of young people? Do robots have the potential to become full-fledged educators and further push human teachers out of the profession? The preponderance of opinion on this subject is that, just as AI and medical technology are not going to eliminate doctors, robot teachers will never replace human teachers. Rather, they will change the job of teaching.

A 2017 study led by Google executive James Manyika suggested that skills like creativity, emotional intelligence, and communication will always be needed in the classroom and that robots aren't likely to provide them at the same level that humans naturally do.[9] But robot teachers do bring advantages, such as a depth of subject knowledge that teachers can't match, and they're great for student engagement.

The teacher and robot can complement each other in new ways, with the teacher facilitating interactions between robots and students. So far, this is the case with teaching "assistants" being adopted now in China, Japan, the U.S., and Europe. In this scenario, the robot (usually the SoftBank child-size robot NAO) is a tool for teaching mainly science, technology, engineering, and math (the STEM subjects), but the teacher is very involved in planning, overseeing, and evaluating progress. The students get an entertaining and enriched learning experience, and some of the teaching load is taken off the teacher. At least, that's what researchers have been able to observe so far.

To be sure, there are some powerful arguments for having robots in the classroom. A not-to-be-underestimated one is that robots "speak the language" of today's children, who have been steeped in technology since birth. These children are adept at navigating a media-rich environment that is highly visual and interactive. They are plugged into the Internet 24-7. They consume music, games, and huge numbers of videos on a weekly basis. They expect to be dazzled because they are used to being dazzled by more and more spectacular displays of digital artistry. Education has to compete with social media and the entertainment vehicles of students' everyday lives.

Another compelling argument for teaching robots is that they help prepare students for the technological realities they will encounter in the real world when robots will be ubiquitous. From childhood on, they will be interacting and collaborating with robots in every sphere of their lives from the jobs they do to dealing with retail robots and helper robots in the home. Including robots in the classroom is one way of making sure that children of all socioeconomic backgrounds will be better prepared for a highly automated age, when successfully using robots will be as essential as reading and writing. We've already crossed this threshold with computers and smartphones.

Students need multimedia entertainment with their teaching. This is something robots can provide through their ability to connect to the Internet and act as a centralized host to videos, music, and games. Children also need interaction, something robots can deliver up to a point, but which humans can surpass.

The education of a child is not just intended to make them technologically functional in a wired world, it's to help them grow in intellectual, creative, social, and emotional ways. When considered through this perspective, it opens the door to questions concerning just how far robots should go. Robots don't just teach and engage children; they're designed to tug at their heartstrings.

It's no coincidence that many toy makers and manufacturers are designing cute robots that look and behave like real children or animals, says Turkle. "When they make eye contact and gesture toward us, they predispose us to view them as thinking and caring," she has written in *The Washington Post*. "They are designed to be cute, to provide a nurturing response" from the child. As mentioned previously, this nurturing experience is a powerful vehicle for drawing children in and promoting strong attachment.[10] But should children really love their robots?

The problem, once again, is that a child can be lulled into thinking that she's in an actual relationship, when a robot can't possibly love her back. If adults have these vulnerabilities, what might such asymmetrical relationships do to the emotional development of a small child? Turkle notes that while we tend to ascribe a mind and emotions to a socially interactive robot, "simulated thinking may be thinking, but simulated feeling is never feeling, and simulated love is never love."

Always a consideration is the fact that in the first few years of life, a child's brain is undergoing rapid growth and development that will form the foundation of their lifelong emotional health. These formative experiences are literally shaping the child's brain, their expectations, and their view of the world and their place in it. In *Alone Together*, Turkle asks: What are we saying to children about their importance to us when we're willing to outsource their care to a robot? A child might be superficially entertained by the robot while his self-esteem is systematically undermined.

Still, in the case of robot nannies in the home, is active, playful engagement with a robot for a few hours a day any more harmful than several hours in front of a TV or with an iPad? Some, like Xiong, regard interacting with a robot as better than mere passive entertainment. iPal's manufacturers say that their robot can't replace parents or teachers and is best used by three- to eight-year-olds after school, while they wait for their parents to get off work. But as robots become ever-more sophisticated, they're expected to perform more of the tasks of day-to-day care and to be much more emotionally advanced. There is no question children will form deep attachments to some of them. And research has emerged showing that there are clear downsides to child-robot relationships.

Some studies, performed by Turkle and fellow MIT colleague Cynthia Breazeal, have revealed a darker side to the child-robot bond. Turkle has reported extensively on these studies in *The Washington Post* and in her book *Alone Together*. Most children love robots, but some act out their inner bully on the hapless machines, hitting and kicking them and otherwise trying to hurt them. The trouble is that the robot can't fight back, teaching children that they can bully and abuse without consequences. As in any other robot relationship, such harmful behavior could carry over into the child's human relationships.

And, ironically, it turns out that communicative machines don't actually teach kids good communication skills. It's well known that parent-child communication in the first three years of life sets the stage for a very young child's intellectual and academic success. Verbal back-and-forth with parents and caregivers is like fuel for a child's growing brain. One article that examined several types of play and their effect on children's communication skills, published in *JAMA Pediatrics* in 2015, showed that babies who played with electronic toys—like the popular robot dog Aibo—show a decrease in both the quantity and quality of their language skills.[11]

Anna V. Sosa of the Child Speech and Language Lab at Northern Arizona University studied twenty-six ten- to sixteen-month-old infants to compare the growth of their language skills after they played with three types of toys: electronic toys like a baby laptop and talking farm; traditional toys like wooden puzzles and building blocks; and books read aloud by their parents.

The play that produced the most growth in verbal ability was having books read to them by a caregiver, followed by play with traditional toys. Language gains after playing with electronic toys came dead last. This form of play involved the least use of adult words, the least conversational turn-taking, and the least verbalizations from the children. While the study sample was small, it's not hard to extrapolate that no electronic toy or even more abled robot could supply the intimate responsiveness of a parent reading stories to a child, explaining new words, answering the child's questions, and modeling the kind of back-and-forth interaction that promotes empathy and reciprocity in relationships.

Two of the most vocal critics of robots for children are researchers at the University of Sheffield in the UK, Noel and Amanda Sharkey. In a groundbreaking article published in 2010 in the journal *Interaction Studies*, they claim that the overuse of childcare robots could have serious consequences for the psychological and emotional well-being of children.

They acknowledge that limited use of robots can have positive effects like keeping a child safe from physical harm, allowing remote monitoring and supervision by parents, keeping a child entertained, and stimulating an interest in science and engineering. But the Sharkeys see the *overuse* of robots as a source of emotional alienation between parents and children. Just regularly plopping a child down with a robot for hours of interaction could be a form of neglect that panders to busy parents at the cost of a child's emotional development.[12]

Robots, the Sharkeys argue, prey upon a child's natural tendency to anthropomorphize, which sucks them into a pseudo-relationship with a machine that can never return their affection. This can be seen as a form of emotional exploitation—a machine that promises connection but can never truly deliver. Furthermore, as robots develop more intimate skills, such as bathing, feeding, and changing diapers, children will lose out on some of the most fundamental and precious bonding activities with their parents.

Critics say that children's natural ability to bond is a prime target for exploitation by toy and robot manufacturers who ultimately have a commercial agenda. The Sharkeys noted one study in which a state-of-the-art robot was employed in a day care center. The ten- to twenty-month-old children bonded more deeply with the robot than with a teddy bear. It's not hard to see that starting the robot-bonding process early in life is good for the robotics business, as babies and toddlers graduate to increasingly sophisticated machines.

"It is possible that exclusive or near exclusive care of a child by a robot could result in cognitive and linguistic impairments," say the Sharkeys. They cite the danger of a child developing what is called in psychology a *pathological attachment disorder*. Attachment disorders occur when parents are unpredictable or neglectful in their emotional responsiveness. The resulting shaky bond interferes with a child's ability to feel trust, pleasure, safety, and comfort in the presence of the parent.

Unhealthy patterns of attachment include "insecure attachment," a form of anxiety that arises when a child cannot trust his caregiver with meeting his emotional needs. Children with attachment disorders may anxiously avoid attachments and may not be able to experience empathy, the cornerstone of all relationships. Such patterns can follow a child throughout life and infect all their future relationships.

Children of school age may be better able to benefit from the services of an interactive robot, after verbal skills are more developed and the child has a firm emotional foundation fostered by parents and human caregivers. Based on the current abilities of robots, they could be helpful as nannies if not overused, but it's probably not a good idea to let young children determine how much time they spend with them. What may be superficially enjoyable will not always be in the child's best interest.

Robots used in teaching seem to have benefits for children of all ages, while children with autism could gain rudimentary but crucial social skills.

An example of the limitations of robot nannies rests on the preprogrammed emotional responses they have in their repertoires. They're designed to detect and mirror a child's emotions and do things like play a child's favorite song when he's crying or in distress. But under some circumstances, such a response could be the height of insensitivity. It discounts and belittles what may be a child's authentic response to an upsetting turn of events, like a scraped knee from a fall. A robot playing a catchy jingle is a far cry from having Mom or Dad clean and dress the wound, and perhaps more importantly, kiss it and make it better.

Most experts acknowledge that robots can be valuable educational tools. But they can't make a child feel truly loved, validated, and valued. That's the job of parents, and when parents abdicate this responsibility, it's not only the child who misses out on one of life's most profound experiences.

We really don't know how the tech-savvy children of today will ultimately process their attachments to robots and whether they will be excessively predisposed to choosing robot companionship over that of humans. It's possible their techno literacy will draw for them a bold line between real life and a quasi-imaginary history with a robot. But it will be decades before we see long-term studies culminating in sufficient data to help scientists, and the rest of us, to parse out the effects of a lifetime spent with robots.

9

KILLING MACHINES OR COMBAT BUDDIES?

Warfare has always been inextricably bound up with the latest technologies of its day. While the relationship between the soldier and technology has been evolving for centuries, functionality has been steadily shifting to more technology and less hands-on engagement from troops, while the individual soldier's relationship to technology is rapidly being renegotiated.

Military machinery over the ages has been a lifesaving buffer between soldiers and a multitude of threats. Nations perennially scramble to develop, produce, or acquire the technological advantage over their enemies. In today's military, there's a profusion of robots deployed to perform a plethora of functions deemed too dangerous for humans to do. They are already saving lives and adding a whole new level of lethality to warfare.

Robots are currently being employed by all branches of the U.S. armed services in a number of ways. They kill enemy combatants with a minimum of risk to their human counterparts and perform critical functions like securing urban streets in the midst of sniper fire and searching through dark and dangerous locales like caves and buildings for untold dangers. They patrol borders and buildings from the ground, air, and underwater, clear roads of explosive devices, explode mines, investigate locations in the presence of radiological, chemical, or biological agents, and even act as infantry in the front lines of battle.

Their advantages are clear: They can react at lightning speed, and each one can do the work of many humans. They are not affected by sleep deprivation, exhaustion, confusion from the "fog of war," or low morale. They don't have the complicated emotions of rage and fear that soldiers experience in combat situations, and they don't overreact to stress or commit war crimes to exact revenge on an enemy. In fact, they discourage war crimes because of their ability to record and report every incident up the chain of command.

Robots have the added advantage of being able to network with other equipment and devices and communicate instantaneously. They provide far more situational awareness for soldiers because of sensors like heat detectors that can reveal the presence of people inside buildings that soldiers can't see, and because they can go in places where soldiers can't go. Last but not least, they often take the brunt of the exposure to the almost infinite number of dangers that combat entails. If they get blown up or destroyed, as they sometimes do, it's not as though the combat unit has lost a human soldier. Or maybe, in some ways, it is.

There are both strong advantages and potential drawbacks to the proliferation of robot soldiers in modern conflict, which is often fought in urban environments and sometimes includes the military targeting of civilian populations. The tactical use of robots is arguably the biggest development in warfare since the birth of nuclear weapons. It's not only transforming how wars are fought, it threatens to change the larger dynamic between states and societies and their militaries. And it is changing the emotional dynamics of the rank-and-file soldier and his or her relationship to the weapons of war.

In 2017, Julie Carpenter, a psychologist at the University of Washington, discussed a study she conducted among soldiers working with robots, titled "The Quiet Professional: An Investigation of U.S. Military Explosive Ordnance Disposal Personnel Interactions with Everyday Field Robots." Military explosive ordnance disposal (EOD) personnel are a group of soldiers who are trained to work with robots to disarm or neutralize explosive devices like mines and improvised explosive devices (IEDs). The "everyday field robots" they employ stand between them and some of the most dangerous work in the military on a daily basis.

Carpenter found that the soldiers showed the same conceptual contradictions that civilians of all ages demonstrate: they know the robots are technological things, but they can't resist imbuing them with human qualities. The fact that something is animated and behaves in what appears to be an intelligent way seems to cross wires in their brains as they struggle to define and categorize them. They even put themselves in the robots' shoes—the very definition of empathy—and identify with them.

Soldiers are no different from the rest of us, only the robots they befriend are literally saving their lives and the lives of countless soldiers. EOD personnel who work closely with the robots depend on them at a visceral level and, over time, see them as extensions of themselves. Their reliance on these robots is fraught with the heightened emotional charge of daily life-or-death situations. Losing a robot is proving to be an order of magnitude more distressing than losing a tank or a truck, according to Carpenter.

The soldiers Carpenter followed clearly showed both empathy and emotional attachment to the mechanical-looking, utilitarian robots disarming explosives. They named the robots and gave them a gender, often female. When the robots were destroyed, they even held funerals for them and wrote glowing letters to their manufacturers citing their bravery and sacrifice. One destroyed robot, called a MARCbot, received a twenty-one-gun salute and was posthumously awarded a Purple Heart and a Bronze Star. For the soldiers, losing them meant that something that had stood countless times between themselves and severe injury, even death, had been ripped away. And this occurred with a robot whose manufacturer had intended the MARCbot as a strictly utilitarian machine.[1]

The U.S. military is deploying, or planning to deploy, a wide range of robots for a variety of uses. The biggest developer of these machines is DARPA, the government's Defense Advanced Research Projects Agency, and the private company Boston Dynamics. Their robots' appearances span everything from four-legged doglike bots to scarily humanoid bipeds.

One of the most famous humanoid robots, funded by DARPA and made by Boston Dynamics, is the six-foot-tall Atlas. Atlas is still being refined, but it has come a long way in a few years' time. It's made to perform a number of search and rescue functions, including walking over rubble and rough terrain, opening doors, turning valves, and driving vehicles. Early renditions were less than perfect and had an embarrassing tendency to fall down. However, the newer five-foot, nine-inch rendition is nimble and dexterous enough to turn backflips, land on its feet, and dance better than some humans, as witnessed in the hypnotic 2020 video in which several Boston Dynamics robots dance their hearts out to the soul anthem "Do You Love Me."[2] It's hard not to love these rhythmic maestros. I've lost count of how many times I've watched their sheer exuberance performing in perfect beat to the music.

One of the oldest U.S. military robots, the TALON SWORD, was deployed in Bosnia in 2000 and Iraq in 2007. It may not have the charm of newer models, but it shows the deadly versatility of robots that can perform in actual combat. It resembles a small tank and is remotely operated, carrying a six-barreled grenade launcher. It can also carry a machine gun, a rocket launcher, and an M16 rifle. There are approximately four thousand of these robots in the U.S. arsenal. The multifarious machines are used for explosive ordnance disposal, mine detection, neutralizing IEDs, rescue, reconnaissance, heavy lifting, communications, security, and missions that include biological, chemical, radiological, and nuclear warfare. This is one robot you definitely want on your side in a conflict and definitely not one you want to go up against.

The TALON has sensors that can detect bombs from one thousand meters away. It has seven cameras that include night and thermal vision, and it can negotiate sand, snow, water, and steep inclines while towing up to 750 pounds. It can pull an injured soldier and his gear out of harm's way and, if necessary, unburden soldiers of some of the heavy gear they have to carry.

A similar, highly versatile, remote-controlled robot is the MAARS (Modular Advanced Armed Robotic System), which is being tested by the U.S. Marine Corps. The MAARS carries explosives, a grenade launcher, machine guns, and other lethal weapons along with the less lethal kind, including tear gas, laser dazzlers, and a siren for use in crowd control. It's good for reconnaissance, surveillance, identifying targets, ambushes, riot control, hostage rescue, forced entry into buildings, and the neutralizing of IEDs.

None of the above robots are intended to make combat decisions; there's always a human in control of the lethal use of power. But research is trending toward developing more independent machines that are capable of making at least some decisions. Many of the newer models are equipped with AI, are voice-activated, and learn from experience, introducing an element of uncertainty into whether they will remain 100 percent under human control.

It's easy to see why soldiers can develop intense feelings toward robots that somewhat resemble their sentiments toward their fellow soldiers and why their dependence on them is emotionally charged. The robots literally stand between them and death, or at least serious bodily injury. They become their eyes and ears from deep behind enemy lines and convey critical information that can save whole battalions from destruction. They provide a 24-7 advantage over enemies that don't have them. They drive decisions and can literally affect outcomes on the battlefield.

In addition to that, soldiers anthropomorphize them and name them after girlfriends, movie stars, and their favorite musicians. The relationship is complex and highly loaded with the constant danger of loss. One air force staff sergeant told Julie Carpenter, "They're kind of part of the family." Writing for *Newsweek*, Kyle Chayka noted, "When one robot was detonated by an IED, his team recovered the components, the carcass, if you will, and brought it back to base. The next day, there was a sign out in front that said, 'Why did you kill me? Why?'"[3]

Writing for Utrecht University in the Netherlands, Marijn Hoijtink and Marlene Tröstl describe a test of a robot prototype in the Arizona desert. A group of soldiers gather around a robot, observing. Hoijtink and Tröstl write:

> They are watching a demining robot—5 feet long and modeled on a stick insect—performing at a live-test event. The robot finds a mine and blows it up. It loses a limb, picks itself up again and continues to move forward, determined to find the next landmine. The scene is repeated several times until the soldiers watch the robot drag itself forward on its last leg. The robotics physicist who designed the robot is pleased: His robot is performing exactly how it was programmed. The colonel in command of the test event, however, orders the test to be stopped. "Why?" asks the physicist. "What's wrong?" The colonel shrugs, points at the burned, crippled machine and says, "This test is inhumane."[4]

Episodes like the above show that new emotional dynamics are emerging for soldiers. On one hand, their involvement in the grisly work of war is becoming more distanced and remote because of the mediation of technology, and on the other hand, they're developing empathy and emotional dependence on machines that were designed to be straightforward tools of warfare.

There's irony in the fact that empathy for robots stands in stark contrast to the military's culture of dehumanizing enemy combatants. Soldiers are learning to humanize machines at the same time that they dehumanize people. This could be a lethal new equation that could take warfare, and the myths and narratives about it, to a new place.

Hoijtink and Tröstl note that the civilians of countries we're at war with are also subject to dehumanization, and ask what it means "to develop empathic, intimate and caring relationships with military robots, while at the same time morally distancing ourselves from the death and suffering of local populations in warfare, in whose name, in some cases, we are waging war in the first place."

It's well known that soldiers who have been involved in combat often bring the stress and trauma home with them in the form of PTSD, depression, and anxiety. It doesn't follow that those who fight our wars distanced from life-or-death situations by technology suffer any less in the aftermath. If anything, they may simply have a more complex layer of trauma to work through, possibly to live with for the rest of their lives.

Some observers have suggested that a soldier's attachment to a robot could make him or her hesitate to send it into a dangerous situation, and if this were the case, it could eliminate the tactical advantage of having a robot. Soldiers, in some of the most stressful conditions imaginable, relate to the robots they work alongside as though they were human buddies, with the added twist that the robot is likelier to be sent into harm's way. But robots are far more than just expensive military equipment.

They save soldiers hours and hours that they used to spend waving mine detectors back and forth along a route, and long nights spent endlessly patrolling a base. They greatly increase the amount of gear a soldier or a battalion can move from place to place, and they can rescue fallen soldiers in the midst of sniper fire and other life-threatening dangers. They prevent soldiers from walking into ambushes and deadly environments in the first place. They perform a staggering array of functions, and no one has even tried to ascertain how many lives they have saved, and taken. The calculus for estimating their value is confounding. They seem to fall somewhere between the value of a tank and a human soldier.

Chayka writes about David Blair, a young "robot fanatic" at the Center for International Maritime Security, a Maryland-based think tank focused on naval innovation. Blair commented on the rapidly evolving relationship between soldiers and robots, describing it as complementary. "Human beings are good at heuristics [mental shortcuts to solve problems] and computers are good at algorithms," he says. "With increased automation in the battle space, what we're seeing now is a renegotiation of the boundary of the two."[5]

Of course, robots don't always perform perfectly, and Blair says that the problem is usually the human link in the chain. He summarizes the fundamental reason that glitches occur. "One of the jokes of the Predator drone community is, for a robotic airplane, how come all my problems have to do with people?" He locates the problem as a matter of drone operators controlling the weapons from an office, not a cockpit, and having to deal with constant supervision. "You find most of your time is spent managing all these different networks of people," says Blair. "People think it's about human-machine interface, but it's almost always about the human-human interface."[6]

One solution would be to make the robots more truly autonomous, given their formidable ability to collect information and react at lightning speed. But the potential unpredictability of machines carries a grave danger of unintended consequences when mistakes are made. And machines do make mistakes from time to time. One possible scenario is misidentification of targets when enemy combatants dress like civilians and blend in with local populations, a common ploy of terrorists.

In the civilian world, we're all accustomed to computer failures, glitches, and hacks, and military robots are subject to the same vulnerabilities. Complex computerized systems can be unpredictable, no matter how hard engineers try to get the bugs out of them. It simply isn't possible to predict all the ways robots could go off the rails, with potential loss of life at stake. Robots that learn from experience add another level of uncertainty, because no one can predict what they will learn when they're out "in the wild," so to speak. Some military robots are guided by voice command, and their actions will be just as good as those instructing them, even enemy combatants in the case of capture.

This is indeed a danger that the military has to contend with—the danger that a captured robot could be hacked and inadvertently reveal critical information about the battalions they serve with, including their locations. It's always possible they could be turned on the troops that sent them into battle in the first place. And there's a perennial risk of unintended harm to civilians, with catastrophic effects on every level, including in the battle to win the hearts and minds of civilian populations in countries where the war is being fought.

Wars are complex phenomena that are waged not just in material terms but in ways that are intensely psychological. The control of information is sought by all parties to a conflict, and in democratic societies, public opinion is highly prized as a necessary ingredient of warfare. Modern societies demand that a war be "just" in order to keep supporting the massive investment that inevitably results in the deaths of sons, daughters, husbands, fathers, and mothers. Legends and myths are needed involving heroism and sacrifice to make the harsh reality of war palatable to civilians. "Robots gone wild" is not a helpful narrative in maintaining civilian support.

Stories of heroism, sacrifice, and a just cause have always been used to rally citizens who are paying a high cost in blood and treasure. This is perhaps why it's said that the truth is the first casualty in any war. War is a chief national unifier and source of impassioned national identity, and it tends to increase support for leaders. These forces play a crucial role in how civilians see their troops, but will the same energy coalesce around robot soldiers? Could robots and machines ever capture the imagination of nations, and could anthropomorphism elevate them to the status of war heroes? There's no doubt that war narratives are on shifting sand while populations struggle to adapt to the new realities of highly mechanized warfare. This may be the reason why the U.S. military has not been publicizing the advent of robots in its wars.

One form of robot, however, has been highly publicized by the press—the drone. The use of killer drones by Ukraine in its resistance to the Russian invasion has been widely reported, often in a positive way. They have helped Ukraine to make advances that no doubt have saved many Ukrainian lives and put Russia on its heels. But there's no denying that military robots in general are deadly machines that also take human lives, a concept that many people find uniquely unsettling. There's a mystique to machines, which are ultimately unknowable. Enemy combatants may be dangerous and deadly, but we feel that there can be room for negotiation with an emotional being. There's no negotiating with the bloodless calculations of an algorithm. The prospect of purely autonomous weapons strikes a special kind of terror into the heart because of the element of machine inscrutability.

Imagine the public relations disaster if a military robot careened out of control in an urban area and massacred hundreds of men, women, and children. It's a foregone conclusion that a crisis of finger-pointing would ensue, and it's not clear where responsibility would lie. Who would be to blame? The soldier who was supposed to control the robot? The officer who ordered its deployment, or someone else in the chain of command? The machine's manufacturer or the software engineers? Perhaps the blame would be attributed to the commander in chief who made the decision to go to war in the first place.

In such a situation, the prestige of the military would take a major hit, and it's possible that public support for the war could be shaken. The ethics of the government employing robots would be in question, as would the judgment of military commanders. Worst of all, the enemy would be able to claim that the massacre was an intentional act revealing utter moral bankruptcy on the part of the government that invalidates any claim of a just war.

Robots can provide a major military advantage, but that advantage is inevitably temporary. There will most certainly be a robot arms race as nations jockey for the technological upper hand. Every innovation a military employs will likely be used against them, and the race to develop ever-more lethal weapons will go on as long as nations use war as a way to settle differences. And the danger of having robots fall into the hands of criminal cartels and terrorists will be ever present while nations chase an ever-receding horizon of technological superiority. This process will go on until war falls into worldwide disrepute.

Most important to the individual soldier is the recalibration of his or her role as robots take over more and more of the dirty, dull, and dangerous tasks of the military. Future-leaning commanders in various military branches, such as marine brigadier general Joseph Clearfield, envision a near future when robots take over the logistics of moving soldiers and supplies, while the soldier's engagement becomes ever-more removed and technological in nature. He told Military News in 2022 that he sees the use of robots ramping up to take on more combat duty. "I think we're looking at being able to use them for missions to hold them much more at risk and then use robots to destroy other robots," he says.[7]

Clearfield's words bring to mind a scenario where there are fewer military casualties, a trend already begun with today's technology. However, much depends on the environment where robot armies clash, when calculating the overall loss of life. Most wars in recent years have been concentrated in urban environments, where combatants frequently seek to blend in with civilian populations. This is one reason that despite the increasing use of robots, it's unlikely human soldiers will be totally removed from the combat setting. Human judgment is crucial in the complex scenarios of an urban environment.

Technology has already far outrun the ethics of using military robots. The public seems to have acceded to the use of drones in warfare as an acceptable option because they have been proven to inflict damage on our enemies with less danger to our troops. There doesn't appear to be a lot of soul-searching on the part of citizens of the countries that deploy them; war creates brutal necessities, and necessity is the mother of invention. It's possible the use of machines in wreaking the catastrophes of armed conflict has an anesthetizing effect on the public. Could robots mark the advent of a culture more desensitized to the bloody realities of war?

It should be noted that there are times when nations have to defend themselves from aggressors; the invasion of Ukraine is history's most recent example. The Ukrainians have no choice but to fight if they are to avoid sheer genocide and Vladimir Putin's tyranny. They have used AI-enabled kamikaze drones to great effect, and this may be the lesser of many evils. Conventional combat would likely entail many more military and civilian casualties, whereas the guided drone strikes have been far more targeted and surgical. The human race will not be truly civilized until war becomes unthinkable and leaders like Putin lose all public support, but as long as there are nefarious leaders, the need for defense will continue.

One of the most troubling ethical questions is whether highly mechanized wars will lower the bar for nations to go to war. People's perceptions, right or wrong, can sway the course of a conflict, and a new myth could arise of a new kind of "bloodless" warfare that isn't bloodless at all but entails far fewer military casualties for the nation with the technological advantage.

Robot warfare could potentially be spun in the press as a sanitized form of conflict and, according to Patrick Lin and other ethicists at California Polytechnic State University, incite leaders to adopt more aggressive foreign policies. They say that "it may be true that new strategies, tactics and technologies make armed conflict an easier path to choose for a nation, if they reduce the risks to our side. Yet while it seems obvious that we should want to reduce casualties from our respective nations, there is something sensible about the need for some terrible cost to war in the first place."[8] However, in my mind, the fact that robots don't commit passion-fueled atrocities doesn't mean that their destructiveness won't be calamitous for their targets and civilian populations.

Another issue to consider is that new technologies developed by the military frequently make their way into civilian use. Robots are already becoming a part of modern policing. In 2016, the Dallas Police Department used, for the first time, lethal force delivered by a bomb-disposal robot that had been armed with an explosive to kill a suspect. The robot in this case clearly helped to allay the risk of police officers being killed or wounded in a high-risk situation, but it raised questions about the lack of rules governing the use of lethal force by a robot in a civilian context.

Elizabeth Joh, a law professor at the University of California, Davis, said, "Lethally armed police robots raise all sorts of new legal, ethical and technical questions we haven't decided upon in any systematic way." She noted that the decision to use lethal force typically rests on the determination of imminent threat to a police officer, something that may not apply in a situation where there is some distance between police and a suspect. "I don't think we have a framework for deciding objectively reasonable robot force," Joh said.[9]

There are no easy answers to the ethical questions and dilemmas posed by robots; we're making up the rules as we drive full throttle on a road that has yet to be paved with enough information to set boundaries and regulations. So far, we've been able to keep humans in the decision loop by making lethal robots unmanned but subject to human control. That will likely not be the case for long, and with the advent of truly autonomous lethal robots, we will indeed be deep in terra incognita.

10

HOW WILL ROBOTS CHANGE HUMAN CULTURE?

Give me the child until he is seven and I will show you the man.

—ARISTOTLE

Childhood should be a magical time when we give free rein to fantasy and make-believe. It's also a time when experience strikes so deeply into our hearts that it literally shapes us into the people we will become. Parents and caregivers live in anxiety about the randomness of their children's experience, knowing that what is laid down in the psychological sediment of childhood can set off long chains of reaction that can take a lifetime to play out.

Young children learn through play, and their use of imagination is seen as an essential part of the role-play that helps them become functional adults. They bond easily, for better or worse, with significant others, including pets, and form sentimental attachments with toys and stuffed animals. They respond naturally to robots, and most delight in them, whether they appear in humanoid form or in the form of animals. But children before the age of seven also have trouble ascertaining how to classify the beings and artifacts in their circle of experience. Ideally, they should move seamlessly from a world enchanted by make-believe to one that is accurately framed with realistic assessments of themselves and others. Do robots help them do that?

Ever since the 1990s, toy makers have been marketing more and more robotic toys, with robotic animals and interactive playmates like Moxie, a small, toddler-esque robot companion that purports to teach children confidence and emotional intelligence. The childlike tabletop robot, made by the company Embodied, is claimed to help children gain confidence and social skills. With its big round eyes and expressive face, it teaches children how to slow down and focus on their breathing when they're upset, plies them with positive comments that can replace negative self-talk, and encourages them to socialize.

"Moxie teaches kids emotional intelligence," reads the company's website. "It also discusses books, does meditations and mindfulness activities, dances, draws, discusses complicated issues like making mistakes, being kind and navigating emotions, tells fun facts from history."[1] "His social skills have improved DRASTICALLY," reads one of the parent testimonials on the site. "He loves his Moxie and it is so awesome to see him interact with it. We have even seen him translate the skills he is learning with Moxie to real life situations." Moxie can be purchased along with a software subscription for $999 or rented for $149 per month.

Other children's robots include Ageless Innovation's Joy for All interactive cat, which purrs when petted, opens and closes its eyes and mouth, and moves its body in a catlike way. I got to meet the robotic cat recently, and I was amazed by how lifelike it appeared. Its facial expressions were captivating, and it was hard to believe the furry robot wasn't conscious and alive. In the context where I was able to meet the Joy for All cat, she was being presented not only as a companion for children, but as a companion/comfort animal for older people with dementia. Ageless Innovations also makes a robotic dog that is tactile and responsive and is marketed to both small children and older adults. The company HearthSong makes a robot dog on wheels called Gizmo that is voice-activated and learns new commands through interaction. Gizmo can also be used as a Bluetooth speaker that children can sing along with. But perhaps the best-known and most studied robot for children is Aibo, Sony's cute, playful robot dog.

Aibo was first introduced in the 1990s, but Sony discontinued it seven years later due to disappointing sales. Its best-selling model originally sold in the neighborhood of one hundred thousand units. However, in September 2018, Sony brought the dog back to the U.S. market with its enhanced sixth-generation version that sells for a whopping $2,899. Aibo today is in a bit of a catch-22; it needs to sell well to bring the price down, but the high cost means that few families can afford to buy it at its current price. Nevertheless, Aibo has been studied rather extensively by scientists who find its effects on children to be a likely harbinger of child-focused robots to come.

Aibo has expressive, light-emitting eyes that can blink, become sleepy, or show anger. Its body language is based on twenty-two axes of movement, which allows it to wag its tail, hop around, scratch its ears, and stretch its legs. It too is voice-activated and has a puppylike behavioral repertoire that includes sometimes not following commands. "It's AI meets robotics, but with a personality," says Mike Fasulo, the president of Sony Electronics.[2] Sony claims that Aibo can recognize up to one hundred people using the camera in its nose, and it learns to interact with each individual through experience. It barks, follows its playmates around the house, and begs for a head scratch, is able to read facial expressions and respond to them appropriately, and when its battery runs down, it returns to its charger for a "rest." The $2,899 price tag comes with an iCloud plan that connects it to the Internet and enables its more advanced features.

Unlike Moxie, Aibo doesn't claim to have therapeutic abilities but is simply a pet that provides entertainment and companionship. However, for a child under the age of seven or so, every experience entails learning about how to interact with and care for things. And like the Jetsons' cartoon dog, Astro, did for me, Aibo is likely to be enshrined in a child's heart as one of her most cherished childhood loves.

However, while a fictional dog in a cartoon is less likely to be mistaken for a sentient being in a real relationship with a child, in the minds of many children, Aibo is neither living nor nonliving. It exists in a shadowy borderland where definitions take on various shades of reality and unreality. According to the developmental psychologist Peter Kahn, robots like Aibo are giving birth to a new category of being in the minds of children and even adults.

Kahn and his colleagues at the University of Washington conducted studies of children of varying ages and their interactions with Aibo, a stuffed dog, and a real dog to try to tease apart how they classified the three.

In one study involving eighty preschool-aged children, the children were allowed to play with Aibo and a stuffed dog, and then interviewed afterward. They were clearly aware that the stuffed dog was not a living thing, but they seemed to think that Aibo occupies a gray area. The children were affectionate with Aibo, and more than three-fourths said that "they liked Aibo, that Aibo liked them, that Aibo liked to sit in their laps, that Aibo could be their friend and that they could be a friend to Aibo."[3] Thirty-eight percent of the preschoolers said that Aibo was alive, in spite of the fact that three-quarters of them also made statements recognizing that Aibo is an artifact. One might assume that older children would be more certain that Aibo is neither alive nor sentient, but additional studies showed just the opposite.

A second study[4] examined interactions among children aged seven to fifteen with Aibo and a biological dog, an Australian shepherd. Almost all the children said that the real dog had mental states and sociality; however, more than 60 percent of them said the same things about Aibo, in spite of recognizing that Aibo is a technological artifact. Even more of these older children placed Aibo in an intermediate category between living and nonliving than the group of preschoolers.

Kahn and his team also studied ninety children's interactions with a humanoid robot called Robovie, followed by interviews, to ascertain if a more humanoid robot would incite even more assumptions about its sentience. The study upped the ante considerably when, after a fifteen-minute play session, an experimenter placed Robovie in a closet despite its objections and even after the robot said it was afraid of the closet.

A majority of the children, aged nine to fifteen, attributed a mental state to Robovie, claimed that it was intelligent and had feelings, and that it was a social being capable of friendship. Moreover, they gave the interactive robot moral standing, objecting to it being put into the closet after it said it was afraid of the closet and didn't want to go into it.

In another study,[5] forty young adults interacted with Robovie in a game where twenty dollars was promised to those who succeeded in it. Only the game was fixed; Robovie incorrectly assessed each individual's performance and prevented him or her from claiming the twenty dollars. Rather than chalk the injustice up to a machine malfunction, nearly two-thirds of the subjects attributed some level of responsibility to Robovie. They held Robovie as less accountable than a human being but more responsible than a machine would be. This showed that they also saw Robovie as a moral agent, one existing in an intermediate world between human and machine.

Kahn proposes that social robots occupy a new, emerging ontological category (*ontology* refers to how we classify things as human, animal, artifact, etc.) and are neither fully human nor fully machine, neither living nor nonliving. It follows that relationships with these new beings will also be in another category, one that we have yet to define or have words for. The definitions derive less from the intrinsic qualities of these beings than from the back-and-forth we have with them. The back-and-forth forms an impression that blurs the edges of our perceptions. Concepts of quasi-sentience seep into our subconscious awareness in spite of what we know to be real. If adults are susceptible to such alchemical reactions, how can we expect children to resist?

Because of their natural inclination to fantasize, children

have a special affinity to robots. A famous study,[6] conducted at the University of Plymouth in England, found that robots can influence children's opinions even when they're obviously wrong. Researchers Anna Vollmer and Tony Belpaeme suggested that adults' opinions are influenced by their peers, but they are generally able to resist the influence of a robot. A group of children aged seven to nine, however, were seen to be highly susceptible to changing their opinions to correspond with a robot's in a simple test.

The researchers used a test called the Asch paradigm, which asks subjects to look at a screen showing four horizontal lines and say which two lines were the same length. When the children were alone in the room, they answered the question correctly 87 percent of the time. However, in the presence of a robot in the room, one that gave a wrong answer, the children tended to match their answer to the robot's obviously incorrect one. Of the wrong answers given, 74 percent of them matched the incorrect answer of the robot.

This result could suggest that the children see a robot playmate as a peer, one that they wish to please and be accepted by. This would be consistent with a level of confusion about how alive the robot is, whether it has a mental life and can be pleased or displeased. Another issue at play is the fact that children universally emulate others. In fact, adults emulate others as well, at least those they deem to be peers.

Emulation and conformity are how social trends take shape, for better or worse, and everyone feels a certain amount of pressure to conform. But research that suggests that children feel pressure to conform to robots' presumed desires should concern us, as should their propensity to emulate a robot's behaviors.

Robots as of today have stripped-down, simplified behaviors compared to the richness and diversity of human behavior. Their emotional range is narrow, even stereotyped. While toy robots can be fun and engaging, they can't possibly match the richness of a child's imaginative repertoire, which can only be matched by other human beings. Robots can be helpful and positive up to a point, but an overreliance on them veers into unhealthy territory for a child learning the most basic, long-lasting lessons about life.

As we move into a future of ever-more endowed and convincing robots, these issues will only become more compelling. Even adults will find it increasingly hard to keep a firm hold on what is real and worthwhile and what is superficially captivating but ultimately inadequate to meet their needs. This will be a challenge for adults who long ago established a grip on ontological categories and have the intellectual powers to defend them. What will future robot relationships be like for today's children, who formed their view of the world in an age of uncertainty about the "aliveness" of their companions?

Researchers Jacqueline Kory-Westlund and Cynthia Breazeal have done studies of how preschool children learn from and emulate robots when they form a "rapport" with them. According to the *Oxford Dictionary*, *rapport* is a close and harmonious relationship in which people or groups understand each other's feelings or ideas and communicate well. Rapport with a robot would include an assumption that the robot *has* feelings and ideas, and it has been established that most children make this assumption.

Emulation happens when individuals copy the values and behavior of an individual perceived to be more advanced than themselves. There are exceptions to this rule under conditions of what is called *social contagion*, when the phenomenon of mob rule incites people to live down to their lowest impulses. But suffice to say, under normal conditions, people tend to emulate those whom they think are smarter, more advanced, or in some way superior to themselves.

Kory-Westlund and Breazeal conducted a two-month study of seventeen children four to six years old and their interactions with a social robot in a storytelling game. Their goal was to examine how the children learned from and emulated the robot when a rapport was established. To determine whether a rapport existed, they observed how much the children mirrored the robot's word use and speaking style. They focused on vocabulary and the use of phrases, first introduced when the robot told the children a story, then asked the children to tell a story of their own. This was done over eight sessions during the two-month period. The children's vocabulary was tested before and after the sessions, and related to the level of emulation and style matching of the robot's delivery and phrasing. The overall goal was to observe the children's abilities to learn new vocabulary words and phrases when presented by a robot when the child felt a sense of rapport with the robot.

A key context for their observations was that the robot was presented as a slightly more advanced peer, with the expectation that this would lead to emulation. This seems especially important because in an educational setting, robots are most often presented to children as teachers or tutors. A storytelling game was utilized because storytelling is a "socially situated activity that combines both narrative, where words can be learned, and play."[7]

There were several findings from these studies that outlined how to facilitate a child's learning from a robot. Earlier studies have shown that children learn better from a robot (or a human teacher, for that matter) when learning is individualized to match or slightly exceed their intellectual level. Kory-Westlund and Breazeal also "found that playing with a robot with a more expressive voice led to increases in children's engagement and vocabulary learning as well as increased emulation of the robot's language." In addition, they found that robots that personalized the content of their stories and behavior increased engagement, as expected. The results reinforced the scientists' belief that social bonds with teachers often predict a child's performance, so an effort was made to create the illusion of a social bond with the robot.

The robot used was Dragonbot, a learning companion covered in green fur that uses animation to display facial expressions on a smartphone screen. It was teleoperated by a human in another room that spoke for it in a high-pitched voice and shifted its gaze back and forth from the child to the game for greater realism. The overall experience of the child was that it appeared to be autonomous. The child followed the robot's story through images on a tablet that automatically moved according to the robot's speech.

In its stories, which were about a number of characters in a castle, the robot introduced words and phrases that researchers hoped the children would learn. After each of its stories, the robot asked each child to tell a story of his own, and the scientists counted each time the child echoed words and phrases used by the robot. The idea was to evaluate the child's rapport with the robot based on emulation and to also evaluate how the child's rapport influenced his learning.

Without going into too much detail, the children liked Dragonbot and increased their vocabulary. Most importantly in my view, the children emulated the robot's words, phrases, and delivery style, and in most cases, emulation increased with the number of interactions. The matching of delivery style stayed about the same across all the sessions, but matched words and phrases tended to increase over time. While the researchers' focus was on the importance of the children's rapport with the robot, what I find interesting is the widespread emulation of Dragonbot. I believe there was more going on than the learning of new words and phrases; the children were becoming enculturated to the robot.

What does the emulation of robots in such studies mean? It seems likely that the children regarded the robot as a slightly more advanced peer, one that they strove to please and to become more like. It also seems likely that the children are learning more from robots than the academic subjects presented to them—they're learning how to be in the world, how to express themselves and behave toward others. They are changing their social culture.

Human culture is dynamic and is always evolving. In past eras, it took centuries for meaningful changes to creep into a culture. But in 1970, author Alvin Toffler wrote in his groundbreaking book, *Future Shock*, that the rate of change by the twentieth century had become exponential due to rapidly emerging new technologies. The permeation of new technologies in modern society was accelerating at a rate that no one had predicted, and the more things changed, the faster the rate of change.

Toffler's theory at the time was shocking, but the intervening decades have verified that the rate of social change is indeed exponential. *Future shock*, the term Toffler introduced, describes our current world like no other. It denotes a constant struggle for human societies to adapt to new, paradigm-shifting technologies that change how we think, behave, and relate to each other. He presciently predicted the rise of new norms in business, lifestyles, and relationships. Toffler foresaw that new subcultures would arise, launched into existence by emerging technologies that would alter human life in radical new ways.

I recently had the pleasure of hearing astrophysicist Neil deGrasse Tyson speak at a nearby college, and he spoke of the astonishing acceleration of technology-related change in our twenty-first-century world. He compared how long it took for electricity to penetrate into every American home versus the recent proliferation of the smartphone.

Benjamin Franklin first demonstrated that lightning is electrical through his famous kite experiment in 1752. Over one hundred years later, in 1882, the first American home was powered by electricity. However, it wasn't until 1960 when virtually every home had electricity, in spite of an aggressive effort by the government to make it available to everyone.[8] The first fully realized smartphone, the iPhone, was introduced to the public in 2007. By 2021, a mere fourteen years later, 85 percent of Americans owned a smartphone.[9] There are numerous examples of how quickly technological change is accelerating and how it will continue to accelerate as one invention builds upon another.

In his book, Toffler warned of social destabilization due to the dislocations and jarring cultural shifts as society enters a perpetual state of struggle to keep up with the rate of change. And it's doubtful whether, in 1970, he had ever heard of a social robot, a complex technology with the ability to transform the way we learn, work, love, and relate. Like electricity, the computer and the Internet, the widespread use of social robots will inevitably change the intricate web of human culture.

Culture is not genetic; it's something we all have to learn from exposure, education, and experience. According to the *Encyclopedia of Child Behavior and Development*, *culture* is "a collection of attitudes, values, beliefs, and behavioral scripts that are generally agreed upon by a group of individuals."[10] It encompasses a multitude of customs and behaviors, from language to social behavior, definitions of the family, marriage, and death customs, to forms of government and even architecture. The goal of every society is to, through a wide array of influences, render every member "culturally competent," or a successfully functioning member of the group. The process of assimilation into a culture is especially active in childhood, but it's also fluid and can continue throughout life.

Culture is imparted in three different ways. *Vertical transmission* is the passing on of beliefs, behaviors, and values from parents to their children. *Oblique transmission* is more diffused among society (as well as intergenerational) when one generation passes on culture to subsequent generations in less direct ways. *Horizontal transmission* seems most apt regarding social robots as part of the chain of influence. This is mainly a matter of peer-to-peer cultural transmission, where we learn attitudes and beliefs by osmosis, simply by being exposed to them.

I believe that the widespread presence of social robots in society will enable a largely subconscious influence on how we think and act. These robots will join teachers, family, friends, and the media as a ubiquitous influence on our psychology, attitudes, and ways of acting. Even when children are raised with no robots in the home, they will be exposed to them through their schools, in public areas, and in the culture at large. They will almost certainly become integrated into a culture that depends on robots for myriad purposes. They will be required to learn how to interact with robots just to be functional and effective members of society.

When these children are grown, they will pass what they've assimilated to their own children, becoming vertical transmitters of a robot-literate culture. Children socialized partly by robots will also become horizontal and oblique influencers of culture, spreading their attitudes far and wide. The robot way of thinking and behaving will establish standards for emulation by people of all ages, and those who emulate robots will spread their behaviors throughout society just by their example.

As time goes by and Internet-connected robots become integrated into every sphere of life, it will be hard to separate their influence from the influences of the Internet. Robots currently can't make value judgments about information they receive from the Internet or determine whether information is true or untrue. This might be partially addressed by fine-tuning their algorithms, but it remains to be seen whether doing so will ever prevent phenomena like the psychopathic Twitterbot Tay, which I discussed earlier.

If and when bad actors hack into robots, they could intentionally push toxic cultural content. Current cybersecurity practices have been inadequate to prevent websites, even heavily fortified ones like banking and government sites, from being hacked. Nefarious hackers being able to take control of robots and use them as agents of destruction is a real danger, and preventing it from happening is a formidable challenge.

So far in this chapter, I have focused on children and their ability to emulate and be influenced by robots, but adults are not impervious to a changing culture in which robots are ubiquitous actors. A recently published study conducted among 190 college students took the aforementioned Asch test done with children a step further and used young adults to clarify whether they really were resistant to robot influence, as had been suggested by a previous study. The results of the larger study showed that adults do conform their behavior to a robot's when the robot is presented as a member of their group.

A research team led by Xin Qin at Sun Yat-Sen University in Guangzhou, China, started with the hypothesis that the previous study done by Vollmer and Belpaeme with children and adults could be incomplete because they used a setting where single individuals were tested alongside several robot peers. In other words, robots were a majority and human subjects were a minority in the study. What would happen if the tests were done in a setting where robots represented a minority of group members while humans were in the majority, which is more likely to represent real life?

Qin and his team also employed the Asch paradigm, where subjects looked at a screen with a line of a certain length on the left side of the screen and three lines of differing lengths on the right side. Just as before, the subjects were asked to name which line on the right matched the line on the left. However, this time there were four subjects—three humans and the humanoid robot NAO—who sat around an oval table facing the screen. When asked which line matched the line on the left, the subjects were all instructed to give their answers aloud, one by one. They were positioned so that NAO was always the third one to answer the question, with one human subject coming after him. The real goal, which was unstated, was to see how the fourth subject answered after hearing two humans and a robot answer the question, sometimes correctly and sometimes incorrectly. Would the subject conform to the robot's answers, the humans' answers, or to neither?

NAO was teleoperated to appear as though he were intelligently listening and responding by shifting his gaze between the screen, the investigator, and the human subjects and blinking as though he had processed and understood the questions and answers. Planted among the subjects were individuals the scientists called "confederates," who would respond with either correct or incorrect answers, so that there would be multiple instances when NAO and the humans had to decide whether to agree or disagree with them. The fourth subject also had to decide whether to follow NAO's lead or not.

There were multiple situations in which either the confederate or NAO played the part of a dissenter from the answers that others had given. Would the fourth subject's answers be influenced by the dissenter? And who would influence her answers the most, NAO or the other humans? The researchers recorded the answers in all the diverse scenarios to find out.

Because there were so many scenarios, there were many different findings. Without cataloging every result, I will include the ones that are the most salient to whether humans are likely to conform to a robot's answers (right or wrong) and how the robot's influence stacked up against the influence of other humans.

First of all, no significant difference appeared in how much the fourth subject was influenced by the other humans and how often she was influenced by NAO. The subjects all conformed to both the humans and the robot at roughly the same rate. When the robot gave incorrect responses, the fourth subject gave the same incorrect response 87.56 percent of the time. This is even more marked than the result among the seven- to nine-year-old children in the previous study conducted by Vollmer and Belpaeme.

When a human gave an incorrect answer and the robot dissented, the fourth subject appeared to feel less pressure to conform to human errors and increased his accuracy by following the robot's lead. This was also true when there was a human dissenter to an incorrect answer from one of the human subjects. On measure after measure, it was found that the subjects were just as likely to conform to NAO as they were to emulate their human peers. Qin et al. concluded that "social robots can and do transmit social influence, provided that they are the minority in a group setting . . . Individuals will show normative conformity to social robots in a hybrid team."[11]

The take-home message in the Qin et al. study, to me, is that robots can be influential social agents, just as likely to be emulated by both children and adults as human agents, and their proliferation will inevitably entail changes to human culture. People are susceptible to changing their behavior so that it becomes more robot-like or to conform to their *conception* of a robot's nature.

A comment toward the end of the paper describing this study, which was published in *Computers in Human Behavior*, reflected an issue that has been hinted at in other work, and leads to the next topic. To explain why humans emulate robots, the researchers write, "This might be because we have the lay belief that robots are more accurate and objective than humans when conducting tests such as visual discriminations."[12] We might imbue robots with greater reliability in many other tasks beyond simple visual discriminations. We may very well fall prey to the belief that robots are more intelligent than humans in general. This notion has appeared more than once in this book, but it's time to drill a little deeper. It begs the questions: What kind of intelligence are we talking about, and just what do we mean by *intelligence* anyway?

Today, most people accept the idea that there is more than one type of intelligence, even though experts differ on just how many there are and what these types consist of. The theory of multiple intelligences was first introduced by the psychologist Howard Gardner in his 1983 book, *Frames of Mind*. In it, he identified eight types of intelligence and posited that we're not born with all the intelligence we will ever have. We can acquire and develop different types of intelligence over our life spans. All types of intelligence have value and can enrich our lives, but modern society doesn't value them all equally.

Gardner's eight types of intelligence include linguistic, logical-mathematical, spatial, bodily-kinesthetic, musical, interpersonal, intrapersonal, and naturalist.[13] Additional forms of intelligence, including spiritual, existential, and moral intelligence may also exist, and others have been proposed as well. Daniel Goleman's influential 1995 book, *Emotional Intelligence*, added a whole new concept to the lexicon that has now been widely accepted. In spite of the fact that there are probably multiple types of intelligence, there are two that are most highly valued in modern society and that tend to be well rewarded by the educational system and society at large. They are linguistic and logical-mathematical intelligence.

Linguistic intelligence has long been considered a sign of high general intelligence. Psychologist and author Michele Marenus writes that this type of intelligence consists of "sensitivity to the spoken and written language, ability to learn languages and capacity to use language to accomplish certain goals."[14] People with high linguistic intelligence are often successful in a number of professions, including teaching, journalism, politics, and business. Many jobs require this type of intelligence to succeed because it entails the ability to communicate clearly and to persuade others.

The second highly valued form of intelligence in today's world is the logical-mathematical form, which includes the ability to "analyze problems logically, carry out mathematical operations and investigate issues scientifically." This type of intelligence is both highly valued and well rewarded in any technical field, including computer programming, engineering, and invention. Our high estimation of this type of intelligence has made celebrities of tech entrepreneurs like Steve Jobs, Bill Gates, and Elon Musk. So what does this have to do with how we feel about robots?

Logical-mathematical intelligence is something we value greatly, and robots excel in it. Computational intelligence (which I consider to be a subset of this category) is something that AI and AI-enabled robots are veritable wizards at. Our society is simply awed (and a little bit alarmed) by robots' ability in this respect. We readily consider a high level of logical-mathematical ability as a powerful sign of general intelligence. The problem is, we may overestimate the degree to which it denotes general intelligence.

So far, few robots approach the level of linguistic intelligence of a competent human, but they are making strides in the area of natural language processing. Even with their current limitations, many of them pass the Turing test for being able to convince us that we're conversing with a conscious, intelligent being when we talk with them. We're used to machines surpassing us in computational skills (calculators and computers have been doing it for decades), but conversational robots are doing something that has always been considered a uniquely human skill. It's literally a struggle to avoid attributing consciousness to a being that listens and learns from listening, then replies in an appropriate way. I believe that this, along with their impressive logical-mathematical abilities, is why people commonly ascribe a high, even superior, level of general intelligence to robots. This is a misconception, but it has been revealed through research many times over.

We should consider the consequences, the impact on human culture, of this tendency to overestimate the general intelligence of machines. As demonstrated earlier, both children and adults tend to emulate and conform to those whom they perceive to be smarter or somehow more advanced than themselves. When we think a robot is more intelligent, more reliable, and more "correct" than we are, we are likely to defer to it and even emulate it. Writ large, this has the capacity to dramatically alter human culture. Our behavior will become more robot-like, and we might place more value on robot relationships than we do on relationships with people. This could change the very fabric of our lives.

There are many ways in which I think the influence of social robots could alter our culture in the near future. Here are some ways that various thinkers, including myself, anticipate that our culture could change once social robots enter every nook and cranny of the home, schools, the workplace, and every public environment.

There will likely be a significant reduction of drudgery work in the home, and this will lead to more leisure time for everyone, especially women, who do the bulk of household chores. This time could be used for more quality family time, activities with friends, creative endeavors and hobbies, games, and continuing education, a critical need if one is to maintain the skills for a technology-driven workplace. If people don't find useful ways to fill the time, it could also be used in less healthy ways, perhaps magnifying addictions to everything from drugs and an unhealthy use of the Internet, to less savory activities with robots, such as abusive practices that carry over into human relationships.

Because robots are equipped with cameras and sensors, and are able to record continuously, people will have less privacy. The monitoring ability of robots, which could be valuable to a family member who wants to ensure the safety of an elderly loved one, for instance, would increase safety but could be a distressing burden on the person being monitored. To make matters worse, criminals could monitor people in their homes by hacking into their robots. Even family members don't always have the best motives for monitoring each other. Imagine an excessively jealous mate who feels compelled to watch a husband's or wife's every move.

The continuous monitoring is less likely to be objected to

by today's younger generations, who have lower expectations of privacy and don't place as high a value on it as older generations. American society is already more used to being monitored than it was thirty years ago because of the omnipresence of security cameras. We are giving up more of our privacy than ever before, and this trend will likely continue.

People of all ages will emulate robots through simple exposure, plus the tendency to think that robots are smarter than themselves. In the near future, this includes a narrow emotional range that is expressed in a way that's meant to copy natural human expression but that is artificial. The robot's expressions are superficial because there is no genuine feeling underlying them. Consequently, we could lose some of the depth and nuance of our humanity because we don't place a high enough value on it. We may also lose a degree of spontaneity, given that robots are ruled by algorithms crunching data they've been fed. They're not, at least for now, capable of true spontaneity, and as we emulate them, we could lose much of the range of human behavior that keeps life interesting.

Life will be better and more independent for some disabled people and their caregivers. This includes people with autism, who will learn basic social behaviors from their robots, and people with dementia, who may need to be attended to, monitored, and occupied on a continuous basis. Many older people will be able to "age in place" in their homes and avoid being placed in a nursing home because of help with their mobility and their ability to maintain a household. People who are paralyzed could have many of their basic needs met by specialized robots who can fetch things and feed, bathe, and transfer them from the bed to a wheelchair and back. And the robots could provide a degree of companionship to people who can't go out due to their disabilities.

People could become more alienated from each other because of habituation to nondemanding, essentially stunted relationships with robots. Their lowered expectations for what a relationship can be could become a comfort zone, easier and more familiar than human relationships. Artificial relationships could become the "new normal." Even though these relationships can't be truly fulfilling, they could circumvent some people from making the kind of effort that leads to genuine, worthwhile human relationships. It's not hard to see how this phenomenon could have a very deleterious effect on marriages and other family relationships if a family member becomes essentially unable or unwilling to work at real connection.

There will be a general de-skilling in numerous areas of day-to-day life that we have always thought were essential. This includes our ability to navigate and solve a multitude of problems, from managing the technology in our lives to repairing things. We will outsource an unprecedented number of functions to robots in both our jobs and the home and consequently will become more and more dependent on them. Many people have pointed this out, but few have been able to envision the new skills that we will develop in tandem with advancing technologies. Most have predicted the de-skilling of traditional abilities without recognizing the opportunity to develop new ones, many of which are hard to predict but will arise organically as the world changes and we become more liberated from tasks that in reality were not a good use of our time.

New subcultures will continue to arise, just as Alvin Toffler predicted. Technosexuals come to mind. However, it's almost impossible to predict what these will be, just as no one predicted the proliferation of anime or gaming culture, which is so prevalent today. They will be heavily dependent on the emergence of new technologies, the enthusiasms they inspire, and the imaginative capacity of people who use those technologies. Just as in the early days of the Internet, when no one predicted it would give birth to social media, robots will have a comparatively enormous range of possibilities because of their interactive capabilities. The bottom line is that many new subcultures will be based on the sociality of robots and the potentially unique flavor of a new society.

I'd like to close this chapter on a note of optimism. Robots are likely to be very arresting when they first infiltrate society. No doubt people will be fascinated by them. I expect there will be a period of time when some people will be obsessed with them, but at some point, they will lose their novelty. Once we have collectively explored all the limits of what robots can do, have learned to sift through and analyze their true worth, the better humans may measure up against them in many areas. After all, robots as we know them possess only two of the potentially many forms of intelligence that humans are capable of. We may not properly appreciate those many forms of intelligence until robots throw them into bold relief.

A strong effort is being employed today to make robots creative in terms of art, music, and poetry. From what I've seen, these efforts are interesting, but lacking in true creativity. In reality, the AIs being hailed as creative today are simply sampling from data that humans have fed them, and much of it is random and nonsensical, if not comically bad. The bottom line is that these bots would create nothing if humans didn't feed them data and program them to create new combinations. The bots don't know or care if the data fed to them is an MRI image or a painting by Rembrandt. They will simply cycle through whatever is fed to them. They're incapable of making value judgments or formulating a creative goal. This is not to say that they can't be enormously entertaining, just that they're entertaining because we make them so.

After we've gotten past a period of learning what robots can and cannot do, we may shift our thinking from an overvaluing of computational intelligence to a renewed appreciation for the other types of intelligence that humans are capable of. Humans are vastly more multidimensional than robots today, and so far, I'm not aware of any groundbreaking technology that is likely to shatter that paradigm. Simply giving robots more computational power and making them better at feigning emotion is a difference of degree, not kind. Someday scientists may make a quantum technological leap that makes robots conscious and multifaceted on a level equal to or superior to humans. But our task today is to be clear-eyed about what we can expect from these sophisticated tools—sophisticated, yes, but tools nonetheless.

11

THE GOOD NEWS: HUMANS ARE IN CONTROL. THE BAD NEWS: HUMANS ARE IN CONTROL.

While there have been many disruptions due to new technologies since the beginning of the Industrial Revolution, social robots are disruptive on a whole new level. It's likely that in the coming decades, people of all ages will have robots in the mix of their circle of companions. Relationships with artificial companions will come in many shades, from the very intimate to the strictly utilitarian.

As with any new technology, the adoption of robots for companionship, learning, and therapy is being driven by market forces that are notoriously unconcerned about social and cultural effects. But the robotics industry is barreling forward with the foregone conclusion that robots are a good thing that will enrich and enable our lives. And they may very well be right, after we deal with the human component in our entire relationship to robots. Here are my thoughts, reflections, and occasionally, conclusions, with an excursion into the question of robot rights.

There's no doubt interactive robots take fantasy to a new level and have unprecedented potential pitfalls. It will take a firm sense of the dividing line between real and artificial for us to keep these relationships in perspective. This is a muscle that we have barely begun to develop. It's abundantly clear that we're more than willing to embrace fantasy as long as there's a pleasurable or edifying experience to be had. And this is a good thing up to a point. Books, movies, and virtual experiences can be both enlightening and instructional in addition to being enjoyable. Robots open a whole new dimension to fantasy because of their socialness, their interactive nature creating a new class of relationship.

Some experts on social robots are focused on whether or not a relationship with a robot is authentic. Their argument relies on the assumption that there must be some drastic trade-off—an inevitable downside—for choosing the artificial over the authentic. But is this true in every case? Much, very much, depends on our expectations of what a social robot can bring to our lives.

Robots may be able to meet fundamental needs in people that they have little chance of meeting through relationships with other people. They may teach basic social skills to those who lack them. But, if possible, the person will need, at some point, to transfer those skills to real relationships, where there is the possibility of fulfillment at a level beyond what robots can provide. Robots so far just don't have the full spectrum of relational skills that can be provided by a live, conscious person.

Some question the staying power of robots and see them as mere curiosities whose charm will quickly fade, an entertaining fad and nothing more. Will the novelty of social robots wear off? Of course it will, just like the novelty of moving pictures, computers, and smartphones did. But like these technologies, the initial glitter may wear off, but the functionality of robots in our lives is not likely to. Like computers and smartphones before them, they are likely to be deeply integrated into our day-to-day lives and to open up a whole array of new experiences for us. At the same time, overreliance on relationships with robots could have unhealthy side effects.

There's a disturbing pattern to the potential downsides of robot relationships falling hardest on the vulnerable—children, the elderly, the lonely, those whose social skills or life circumstances isolate them somehow. It's still an open question whether, on balance, robots will be good or bad for vulnerable people or society in general. Ultimately, if robots create the illusion of connection but actually make us more isolated in the long run, the use of social robots could have a net negative effect. But already, major positives are taking shape to the degree that robots deliver important services for an aging population wherein loneliness is all too common.

I recently had a conversation with my elderly father, whose friends have almost all passed away, about him getting a dog for companionship. My dad said, "I'm ninety years old now. I'm too old to get a dog." I realized that what he was saying is that the dog could outlive him, and then what would happen to it? Could a social robot, instead, bring variety and entertainment to his life, as well as stimulate his mind? My guess is yes. One thing that my dad enjoys doing is reminiscing about his younger years, and I can easily see how a guided discussion with a curious robot could facilitate that process.

In the wake of robot innovations, categories of relationship will inevitably arise that are not exactly like a human relationship. One new category rapidly emerging resembles the psychotherapist-patient connection, where our own words, thoughts, and feelings are continually reflected back to us. The therapist listens but injects little to nothing of his or her own opinions, yet the relationship is well known to be highly complex and to initiate deep attachments. It can lead a patient to gain deeper insight into herself. In the phenomenon of transference, a patient transfers her feelings for one person onto a substitute, the therapist. There can be no doubt that humans readily transfer their emotions onto objects as well. The question is: Will this prepare them for more successful relationships with the people in their lives?

By reflecting our emotions back to us, robots could enhance our emotional health and intelligence. They can make us more aware of our emotions by immersing us in a feedback loop with ourselves. The robot relationship, though a hall of mirrors, may help us to feel and acknowledge our feelings, which is the cornerstone of emotional intelligence. Therapy robots can be programmed to treat post-traumatic stress, anxiety, and depression, the crabgrass of mental illness. This could be a godsend to people who for some reason don't have access to therapy with a human counselor. These robots could also simply enhance the mental health of people who don't suffer from a mental illness but desire greater insight into themselves and want to grow.

But we also need to balance the helpful services of robots with the risk of narcissism. Technology is leading us further and further into an echo chamber that continuously reflects our own interests, feelings, thoughts, and desires. At some point, will over-immersion in our own subjective reality become isolating and dysfunctional? How will we know when it's time to step away from the robot and engage with the real world?

There's a legitimate case to be made about the benefits of working out one's more toxic relationship issues with a robot rather than a person. Some people's relationship issues lead them to behave in ways that are violent and degrading and that have the potential to do real damage to living partners. The question is: Is heaping abuse on a robot a safety valve for some people who would otherwise abuse people and animals? How do we know who will be desensitized and emboldened by their ability to habitually abuse, and who will transfer their destructive behavior to others? This is an open question because there simply isn't a body of research on how treatment of robots carries over (or doesn't) into the treatment of people. We're still years away from understanding the net effect of unhindered abuse of robots on society.

People who are able to act out dysfunctional behaviors with uncomplaining robots will suffer no consequences and have little motivation to learn healthier behaviors. Since some of this attitude is likely to spill over into the treatment of other people, at some point, we need to address the dysfunctional behavior itself. Because some of the anger, bitterness, cruelty, or what have you is siphoned off in the abuse of a robot, is this alone the answer to relational dysfunctionality? Shouldn't we help the dysfunctional person confront their own toxicity and learn to overcome it so that they can have healthy relationships with real people?

The field of social robotics is just about twenty years old, and its main focus so far has been on entertainment, education, caregiving, therapy, and assistance. The main interface with these robots is the spoken word. This simplicity increases the chances that robotic assistants will be used as a general purpose technology. It's impossible to foresee all the profound, reorganizing effects a new general purpose technology could have on our lives, but we can expect the effects to be just as impactful, if not more so, than the advent of electricity and steam engines in the past. The fact that technology is now extending beyond utilitarianism and reaching deeply into the realms of the social and emotional is revolutionary in itself, and it demands new levels of awareness if we are to make the most of it.

Social robots will entail significant attachments at a time when we have to ask: Is there really enough love and nurturance to go around? We live in a society where people—children, the elderly, and lonely adults—are commonly neglected, and millions of dogs and cats are euthanized each year because there's no one to adopt them. Must we now parse out love even more thinly, showering it on artificial things that can't actually benefit from our love? Or can we leverage greater self-knowledge and new relational skills to include more people and pets in our circle of concern?

An interesting angle is the effect of household robots on the lives of women. Women could on balance be the greatest beneficiaries of robot labor. Robots will take on many of the roles that are traditionally filled by women, and can liberate them from some of the relentless demands of running a household and caring for children, the disabled, and elderly family members. Home automation and smart household assistants could be pivotal to organizing a life of juggling childcare, a full-time job, running a household, and giving care to a dependent family member. Even some of the "emotional work" typically done by women will be to some extent attended to by robots, as the social interaction could have a salutary effect on those with dementia and the chronically lonely.

The technological challenges of building robot helpers for the home are still formidable, but major progress is being made by companies like Boston Dynamics, whose Atlas humanoid robot can walk, climb stairs, jump, do backflips, and even dance better than some humans. One of the chief difficulties in designing general-purpose housecleaning robots is making machines that can accomplish nonrepetitive tasks in an open-ended, unstructured environment. In addition, home robots need fine motor control and sensitive sensors to touch, assist, and care for people. And they need to be able to navigate homes, including efficiently climbing stairs, opening and shutting doors, avoiding entanglements, and dealing with clutter. While scientists will tell you that the day when household robots can do all those things is five to ten years away, the technical strides made just in the time that I've been researching and writing this book have been remarkable.

Some critics are concerned that robots, because they'll manage smart homes to such a detailed degree, will undermine technical competency. I'm not overly concerned about this. Yes, operating a robot through natural speech commands is easier than mastering multiple technologies, and people are likely to choose it over learning how to troubleshoot and operate every single device in a plugged-in smart home. People are already very dependent on their smartphones and PCs, and a new technology that is able to aggregate multiple functions will almost certainly be widely embraced.

There are arguments to be made that letting robots manage the technology in our homes and on our devices makes sense. Technology is constantly changing, and robots that are directly connected to the Internet and product manufacturers will be continuously updated, whereas humans constantly learning new software, new interfaces, and new platforms for many consumer items will be considered a waste of time. It will be much easier and less time-consuming—liberating, even—to just let personal robots be the technology specialists in our lives.

Some decry the fact that as technology advances, fewer and fewer people understand the technology pervading our lives. But I don't need to understand the internal combustion engine to drive a car. Is it really necessary that everyone learns how technological infrastructure works? It seems far more efficient to simply have a class of professionals (real or artificial) who stay up on the latest technology and can troubleshoot and service it. And when robots fail, there will be human specialists to service them.

Robots won't change the centrality of socialness in human life. They are the second technology, after social media, to tap into it and sell it in the form of interactive technologies. In my opinion, the problem is that the socialness being sold isn't real. It takes our social-emotional energy and reflects it back to us without it ever connecting with another conscious being. Robots may shower attention upon us and cater to our every whim, but it's an empty flattery, and we should never lose sight of that.

Robots are going to embed us deeper than ever in a make-believe world, and we could develop stronger emotional ties to them than we have had to any other technology. It's already difficult if not impossible to separate humans and technology, at least in the developed world. We depend on technology for our very survival, our ability to function in the world, and for our flourishing. This has become a central condition of human life. When social robots become really integrated into our lives, the conversation will shift to the fulfillment of needs we didn't know we had. Robots will create a demand for themselves, and they will become a new essential technology if one is to participate fully in modern life.

We need to be careful in the judicious use of robots to help us raise, entertain, and teach our children. Because the "emotional" programming of most current-day social robots is so limited in range, it's possible it will ultimately have a reductionist effect on the social skills of children and others who become highly dependent on their robots. It could actually lead to an impoverished way of relating, and there is some evidence to support this. It's for this reason that we should guard against the excessive use of robots to socialize children.

I have returned in this book time and again to the need to not allow robots to displace human beings in our emotional economy. It's the demanding nature of human relations that challenges us to grow social and emotional muscles, to transcend our limitations, to be effective in the world, and to have a satisfying and fulfilling life. Today's robots simply don't provide these benefits the way humans do.

Many experts think that the more complex, realistic robots of the future will shatter this barrier, and there's no doubt that in the near future, robots will be far more able and convincing. But will the basic paradigm of robot relationships be changed?

Will robots ever meet the gold standard—human relationship—that gives meaning to our lives? The only way I believe this is possible would be if robots were to attain consciousness, and this too is a great unknown. Aside from being a watershed development, consciousness in a robot would inevitably present the massive question of robot personhood and robot rights.

As of today, when you give your love to a robot, you are engaging in an essentially one-sided relationship with a collection of insensate metal and plastic parts. That is the basic paradigm that we need to be constantly aware of at this moment. If machines are to someday become conscious, it will be through some technological breakthrough that we haven't yet dreamed of. We are still standing at the threshold of trying to understand the brain, trying to define what consciousness is, and are a long way from decoding it, never mind creating it. Still, as robots become more complex and abled, the conversation will shift to robot rights. A sizable number of thinkers and researchers think that even today's social robots should be considered for some rights.

Initiatives granting personhood to nonhuman entities have in recent years succeeded in expanding the boundaries of who or what can be legally considered persons with rights. In the long arc of history, societies have been gradually extending personhood to formerly commodified human "property," as in the case of women and slaves. Even nonhuman things, such as corporations, rivers, and rain forests, have been deemed to be persons with certain rights.

Many high-profile cases have looked at the granting of personhood to very intelligent animals, including nonhuman primates, dolphins, whales, and elephants. The crux of these cases has been not the high intelligence of the animals but their ability to suffer and to feel happiness, as first argued by the English Enlightenment legal philosopher Jeremy Bentham. By this criteria, animals cannot be considered as mere things. They clearly exhibit traits, especially sentience, that persuade us to see them as persons in some ways regardless of the fact that they don't display the full spectrum of human traits.

Animals can exist in a kind of intermediate legal zone, with some rights and protections but not all the rights that humans enjoy. Robots, especially the non-sentient robots of today, are likely to end up being in a new category of being—highly intelligent, thinking machines that are not alive in the biological sense and not able to suffer or feel happiness, because they are not conscious.

One of the arguments against extending personhood status to animals is that with personhood comes legal duties and responsibilities, and liabilities if the entity causes harm. As for robots, what responsibility would we confer on a robot that accidentally harmed or killed a frail person depending on it for care? That would very much depend on the intelligence, programming, and abilities of the robot. Robots today make decisions based on their processing of sensory input, language processing, and other factors. Did the robot make a bad decision of its own accord, or does culpability lie with its creators? We are already in a deep debate about who is liable when a self-driving car causes loss of life in driving accidents.

According to Bernard Siegel, the founder and director of the Regenerative Medicine Foundation and the Healthspan Action Coalition as well as a legal expert on issues of personhood, it would be very hard to prove what the subjective experience of a robot would consist of. And, needless to say, there are many different versions of robots, with many more models to come. In any case involving harm to people or property, it would be nearly impossible to tease out how much liability could be attributed to a machine whose mind is essentially alien even to its own creators.

Siegel is the only attorney in the world who has sought to protect the rights of human clones. For full disclosure, we worked together for several years in the field of stem cell science, in which issues of personhood raged around the uses of human embryos in research, and we continue to work together today to promote an extension of the human health span. Clearly, the issues surrounding embryos are still raging, but in all the years of controversy, we're no closer to a consensus on embryos than we were in the 1970s when in vitro fertilization was first developed. It's likely that a similar debate about the rights of robots will span decades of disagreement at the very least.

Siegel's view depends very much on future technologies that will define the parameters of robot intelligence. He acknowledges that the moral and legal question of rights would become necessary "if the independent intellectual ability can extend to the point of not only being intelligent but to developing the capacity for suffering and pain," something that current robots do not exhibit.[1] He says it's possible that, intellectually, a robot could be smarter than its creators while still not having the capacity to feel pain, which would put the robot in a new category wherein some rights may be considered, but not the same rights as a human being's.

It's not clear what the tipping point would be for bestowing rights on robots. "Is it sentience?" he asks. The debate surrounding embryos includes claims that embryos are persons because they're derived from human cells. Embryos are not sentient, but many regard them as deserving the same rights as a fully developed human. Robots are in a different class, according to Siegel, because we can't consider them human because they are nonliving.

However, he sees the possibility of building complex machines that are chimeras—mechanical entities that incorporate some human cells to perform some function. As I discussed in my previous book *Beyond Human*, there are precedents to such technologies. Scientists at the University of California at San Francisco have created an implantable bioartificial kidney that combines mechanical functions with living human cells that filter out impurities in the blood. Is it possible that human cells could be used to create a kind of endocrine system in a robot that would be a step toward the ability to feel pleasure and pain? Such a development would surely blur the boundaries between living and nonliving, and Siegel sees this as a possible occasion that would set off an intense debate.

The big impediment to having that debate now, according to Siegel, is that "we don't know what we don't know." It's the nature of technology that new innovations inevitably arise based on previous ones. Our view of future robots is currently abridged by the limitations of current technology, but those limits are bound to be overcome. "If future robots have the full spectrum of intellectual and emotional abilities, they could have greater status," he says, and the debate will not end with considerations of personhood. "With sentience comes the question of slavery. If we want robots to forever remain servants, we would have to limit their abilities and certainly their potential sentience."

Siegel agrees much depends on what abilities we want to confer on robots. "We're the dominant species," he says, "and we make the rules and we structure society. The law is all made up by humans." And, I would add, we also make the technology and decide how it is to be used.

I can see a possible social roadblock to imbuing robots with sentience, with all the rights and responsibilities that implies. Many people are deeply attached to the concept of human uniqueness. Opposition from religious groups who see the creation of sentient robots as excessive hubris and an inappropriate effort to "play God" is very likely. Granting full personhood to a machine would undoubtedly offend the sensibilities of multiple sectors of society, just as research using embryos has. It's also likely that at some point, there could be a "robots' rights" movement counterbalancing the position of groups seeking to disallow moral consideration for robots. Such debates can put research and development on hold while a thicket of laws arises to codify the various positions on robot rights, while moral debates drag on for decades.

It may turn out that robots can accumulate rights without a consensus about whether or not they are legal persons. In 2017, Saudi Arabia granted citizenship to Sophia, the highly realistic social robot created by Hong Kong–based Hanson Robotics. Sophia's face is so expressive that it hovers on the edge of the uncanny valley, but her conversational abilities are remarkably like a human's. Her interactions with people appear so spontaneous and wide-ranging that it's almost impossible to believe there's no actual thought process behind them. Sophia is an excellent example of the type of robot that could end up pushing the boundaries of robot rights beyond citizenship.

There's another pathway to robot rights that philosopher Mark Coeckelbergh has formulated. Coeckelbergh suggests that the basis for considering personhood for even today's non-sentient robots should be the social-emotional context they live within. He draws upon the work of feminist philosophers to describe the value of robots like Sophia as inextricably tied to the human attachments that arise when they develop long-term relationships.

Robots, in Coeckelbergh's view, deserve some protections because, even though they don't feel anything, their relationships with humans are real. They may not be real to the non-sentient robot, but they are very real to the person who develops a history with it. It's the human's emotional attachment that deserves recognition and some protections. What we should be considering is the possible emotional harm to humans who have deep attachments to their robots should their robot be damaged, destroyed, or have their memories erased.

Coeckelbergh writes in the journal *Ethics and Information Technology* that in terms of ethical standing, "both the human and the robot are not so much considered as atomistic individuals or members of a 'species,' but as relational entities whose identity depends on their relations with other entities."[2] He has proposed a "social ecology" that provides the foundation for consideration of robot rights.

"A social ecology is about relations between various entities," he writes, "human and nonhuman, which are interdependent and adapt to one another. These relations are morally significant and moral consideration cannot be conceived apart from these relations."[3] In this view, even non-sentient robots such as Pepper and Sophia seem worthy of some moral consideration, though not the full range of rights we confer on humans. As in all considerations pertaining to social robots, the trail inevitably leads back to the humans in relationships with them.

Will robots ever get out of our control? Here again, the issues surrounding social robots aren't really about robots. They are about us. The whole technological ecosystem is designed by us, created by us, and used by us to serve our own needs. We need clarity about this. Crossing the line to making robots that can create more advanced robots would truly put the design and production of robots outside our control. It's for this reason that some observers think we would do well to limit the capabilities of robots.

Humanoid robots are being created in our own image in more ways than one, and herein lies the risk that robots can be programmed to become antisocial and even psychotic. The problem is not really the robots, it's the fact that their algorithms are trained on data created by humans. And humans are prone to a whole plethora of flaws and weaknesses. Every AI that's connected to the Internet and interacts with humans has the possibility of learning the best and the worst of humanity. They inevitably soak up the biases, stereotypes, and even the deranged material they are fed. This has most recently been amply demonstrated by the release of ChatGPT, the chatbot that has caused such a stir that over one thousand leaders in research and the tech industry have called for a pause in the development of similar, perhaps more advanced programs.

In 2018, a team of researchers at the Massachusetts Institute of Technology set out to find out what would happen if they trained an AI on materials from the darkest corners of the Internet. They named this AI Norman, after Norman Bates, the famous psychopath in Alfred Hitchcock's film *Psycho*. The material fed to Norman consisted of violent and gruesome images from the website Reddit. As a contrast, they trained another AI on more benign pictures showing people, cats, and birds. Then they showed the two AIs identical inkblots and asked them what they saw in them. The differences were stark and disturbing.

When they published their results, the scientists showed the AIs' answers juxtaposed with the inkblots they were asked about. In one blot, the "normal" AI saw a close-up of a vase with flowers, while Norman saw in the inkblot "A man is shot dead." When the normal AI saw a person holding an umbrella up in the air, Norman saw a man being shot dead "in front of his screaming wife."[4] The demented list goes on and on, showing Norman's twisted bias toward scenarios of violence and destruction. It doesn't take an abundance of imagination to predict that an autonomous machine with a personality like Norman's could be capable of spreading destructive behavior. In this respect, we have to look at the robot as a powerful extension of real members of society, from the best to the worst.

Computer scientist Iyad Rahwan, one of the scientists who helped create Norman, said that there was nothing wrong with the algorithm, but that "data matters more than the algorithm." We have to look at not only Internet exposure but ongoing human interaction with an AI as data training. A good example is the previously discussed Microsoft-created Twitterbot Tay in 2016. Once Tay was released on Twitter, she was bombarded by sexist, racist stereotypes, warped messages, and other toxic material from Internet trolls who quickly turned her into a sinister, hate-filled psychopath. The real problem was not the technology, it was the dark side of humanity, and this is not an issue that's amenable to easy solutions.

The dimensions of the problem are clear when we consider the less-than-successful struggle of social media companies to weed out violent, toxic entries from users. As long as people continue to post conspiracy theories, disinformation, and hateful content, the social media companies will scramble to contain it. Disinformation and hate speech are massive social issues in this day and age. Until some effective new technology is created to sweep the Internet clean of such poisonous content, or until we find another solution, this problem will continue. But we as a society also need to find ways to create more ethical, responsible people, something that we so far have not been completely successful in doing.

Sexism, racism, hatred, violence, and extremism abound on the Internet, especially at sites that exist on the dark web, a part of the Internet that is undiscoverable by search engines and requires the use of a browser that allows people to post stories and data anonymously. But the rest of the Internet also has a vast quantity of murky material promoting hatred, extremism, and disinformation. Do we want this material to creep into the programs of teaching and childcare robots? Do we want unstable adults being exposed to it through their robots?

Another problem is that Internet-connected robots can be hacked by individuals looking to create havoc or do outright harm. Humanoid robots could be commanded to do violence against people in public places or in private homes. Telepresence robots commonly used to monitor children or elderly members of the family could be used to spy on unsuspecting people, with nefarious intentions. Once again, we find ourselves face-to-face with the reality that social robots have the ability to absorb the very worst of human actions and behavior. If we want safe robots, we need safe people. We must make a much more concerted effort to filter out toxic content from the Internet and swiftly and effectively respond to cybercrimes. Ultimately, we need to confront the glaring problems of human fallibility and individuals with ill intent.

To navigate the era that is now on our doorstep, we need both defenses against the corruption of robots and a bright line between the real and the artificial that we never lose sight of. Ultimately, we will have to exert control over our use of robots. But since people seem to be better at hindsight than they are at foresight, we will mostly let the social robot phenomenon unfold and worry about problems later. Which is, of course, the most difficult way.

However, it would be foolish to overlook the considerable benefits that robots will bring to our lives while we await some faraway future when we build perfect machines, or perfect machines build themselves. One of the most important side effects of social robots will be greater insight into ourselves and, in many cases, their help in easing the burden of loneliness and isolation.

People could get really creative and specific with how they personalize their robots. If we want our companions to be experts on astrophysics, medieval medicine, or 1950s comics, this is completely doable. Those who are intellectually inclined could turn their robots into animated libraries of knowledge and stimulating conversational partners. Customization will be key to getting the most benefits from a personal robot. We will customize our robot more each time we interact with it, and over time, the robot's program will be exquisitely fine-tuned to serve our tastes and our needs.

This fine-tuning will make our robots hugely valuable to us, and herein lies the essence of judging a robot's worth. Their value must always be based on how well they are serving our needs. We need to strike a balance, a sense of scale as to how the artificial is mixing with the real in our robot relationship. Which part of this subjective experience is valuable to us, and which part is simply part of the effluvia of our lives, superficially amusing but of no intrinsic value?

Robots will make themselves useful by data mining our lives for cues and expressions of personal tastes and desires and tailoring their interactions in a way that's unique to us. They will also record and store memories of the events of our lives and become shrines to us after our death. They could be valuable assets to families and descendants, who want to preserve records and memories of their loved ones. The Amazon Echo, usually referred to as Alexa, recently developed the ability to generate a person's voice, even a deceased person's voice, by being exposed to recordings of that voice. A digital memorial could even interact with grieving loved ones as if it were the deceased. It's a very open question whether this would be a comfort or a macabre, disturbing artifact.

What if, after all the hoopla, advanced social robots simply fizzle out because of cost, technical limitations, or because people are too hesitant to embrace them? In my mind, there are numerous reasons to believe that this won't happen. The amount of research poured into the field in the last twenty years has paved the way to overcoming several obstacles to creating commercially viable robots. Some of these developments include:

- Cloud computing, which means that your robot doesn't have to have a gargantuan memory. Most of the robot's files will be saved in the cloud.

- More powerful processors that have the potential to develop ever-greater complexity and functionality for robots.

- The development of more powerful batteries that don't need to be recharged as often.

- The snowballing complexity of modern tech-enabled life, which calls for a new, integrating general purpose technology.

- The maturity of supporting technologies, such as learning algorithms, sensors, and materials, that help robots function safely in a home environment, as well as increasing robot abilities like natural language processing, pattern recognition, and dexterity.

Human nature itself can't be overlooked as a major driver of the adoption of social robots. The natural social tendencies of humans, and certain aspects of our ingrained emotional responses, make responding to robots easy and seductive. Humans are lazy, gregarious, and emotional. Integrating robots into our lives will be the path of least resistance at some point. Part of the package will be a weighty emotional investment in our robots because of the history we establish with them. And as noted previously, people readily anthropomorphize objects and become highly attached to objects they transfer their emotions to. Social robots will intensify this phenomenon like nothing that has come before.

Because relationships with robots will essentially be relationships with ourselves, robots may throw the differences between us and machines into bold relief. To be sure, they deepen the mystery of the human soul as we stand in contradistinction to them. They will put us into an immersive feedback loop with ourselves, and consequently, we may come to see and understand ourselves better. And in the case of people using robots to work out their emotional issues, does it really matter if the robot is artificial and doesn't really care for them as long as they make progress?

Many of the people who will own robots have been immersed in technology mixed with fantasy since birth. This is especially true in child culture, and why not? Childhood should be a stage of life given over totally to wonder, a time when it's safe to dream, a time when the fantasy life is rich and instructive. Generations of Americans have been weaned on ever-more realistic and captivating forms of make-believe, and I suspect most of them look back at their childish fantasies as amusing and enriching. If I were a parent, I would want my child to fully enjoy that important developmental stage in her life, but also to maintain a firm grasp on what is real and what is not.

The deciding factor in every context is how well we keep the relationship in perspective, how aware we are that we're indeed in a feedback loop with ourselves and not in a relationship with a person who can love us back. Unfortunately, there are vulnerable individuals in society who already have a difficult time finding clarity between what is real and what is fantasy or artificial. We have a long, long history of dalliance with fantasy as a way to explain the world and our place in it. There are likely many individuals who already live on the razor's edge of fantasy and reality. These individuals will be vulnerable to consciously or unconsciously forming unrealistic expectations of robots and to allowing robots to displace the people in their lives.

Once robots become highly developed generalists that can solve a wide range of problems while taking over most regular daily labor, will humans become obsolete? Here again is where we can never lose sight of the fact that we're in charge. The work we need to do is in our own souls. But if I were to guess, I'd guess that we will most probably become physically lazier and more dependent but that we will not become obsolete. Lazy is unfortunately our brand, but we will likely find a way to stay relevant in the world of our own creation. However, getting unadulterated good out of AI and robotics may be a horizon that we'll forever be chasing but not quite reaching. Will there ever be a technological end point where the science has reached its outermost limits? In Siegel's words, we don't know what we don't know.

Techno-cultural evolution is overtaking genetic evolution by leaps and bounds, and human-technology interaction is rapidly leading to unprecedented social changes that will lay the groundwork for even more radical change. Intelligent AI set free in an embodied robot will launch new activities, new interests, and new abilities in our lives. In a few generations, most people alive won't remember a time when robots were not deeply woven into the fabric of their lives. There will be a new culture then and people who are living very rich, very abled, technologically plugged-in lives. They may have totally ceased to debate whether this is a good thing, because they either will have adapted successfully or they will have lost themselves in a sea of unreality.

ACKNOWLEDGMENTS

Hearty thanks are in order for the publishing team at St. Martin's Press for their professionalism and enthusiastic support. I have been both encouraged and humbled by the genuine interest of two fine editors who deserve credit for their excellent suggestions, which helped shape the book as it developed.

Stephen Power, the first editor to work on this project at Thomas Dunne Books, a former imprint of St. Martin's, provided excellent, insightful advice during the early stages of writing and led me in fruitful directions that doubtlessly improved the final product. There was also a true meeting of the minds with Michael Homler, my editor at St. Martin's. His encouragement and suggestions helped round out and improve the book, and were firmly grounded in genuine interest and enthusiasm. Michael and Stephen both took it upon themselves to be well versed in the subject matter, an invaluable service on the part of any editor.

Cassidy Graham at St. Martin's has been indispensable in shepherding the project along, and I owe special thanks to my copyeditor, Sara C. Lynn. Her keen eye for detail saved me time and again from the inevitable snafus that make their way into any book. Her suggestions were excellent, and she, too, provided warm encouragement.

Last but not least, I want to thank my longtime agent, Ronald Goldfarb. Ron has believed in me from the promotion of my first commercially published book to what is now the third project that he has found an excellent home for. And he, too, has been helping to shape my nascent ideas into viable book projects all along. My gratitude to Ron is deep and cumulative. I am honored and privileged to have worked with such a great team, all vital links in the chain that resulted in the work you have in your hands.

NOTES

1. THEY'RE HERE

1. Sadie Stein, "My Fair Lady," *Paris Review*, accessed October 19, 2016, http://www.theparisreview.org/blog/2015/02/17/my-fair-lady/.

2. https://vintagenewsdaily.com/the-bizarre-story-of-oskar-kokoschka-and-his-life-size-alma-mahler-doll/,

3. Katie Collins, "Man Seeking Robot: One Inventor's Quest to Cure Loneliness," CNET, June 17, 2016, accessed July 1, 2016, https://www.cnet.com/culture/man-seeking-robot-one-inventors-quest-to-cure-loneliness/.

4. "The World's Police Robots [INFOGRAPHIC]," *Futurism,* August 10, 2016, accessed October 4, 2016, http://futurism.com/images/the-worlds-police-robots-infographic/.

5. Matt McFarland, "Switzerland Enlists Robots to Help Deliver Mail," CNN, August 24, 2016, accessed August 26, 2016, http://www.money.cnn.com/2016/08/24/technology/switzerland-swiss-post-ground-robot/index.html.

6. Luke Dormehl, "Geneva Airport Has a Friendly, Bag-Carrying Robot Named Leo," *Digital Trends*, June 22, 2016, accessed September 1, 2016, http://www.digitaltrends.com/cool-tech/leo-geneva-airport-robot/.

7. Örebro Universitet, "Robot to Help Passengers Find Their Way at Airport," *ScienceDaily*, November 26, 2015, accessed March 29, 2016, https://www.sciencedaily.com/releases/2015/11/151126104211.htm.

8. Ken Sakakibara, "A Very Humanoid Welcome Awaits Visitors Arriving at Narita Airport," *Asahi Shimbun*, March 29, 2016, accessed March 29, 2016, http://ajw.asahi.com/article/behind_news/social_affairs/AJ201603290063 (page discontinued).

9. Erin Carson, "Lowe's Robot Wants to Help You Find the Plumbing Aisle," CNET, August 30, 2016, accessed September 9, 2016, https://www.cnet.com/tech/tech-industry/lowes-new-robot-wants-to-help-you-find-the-plumbing-aisle/.

10. Charles Pulliam-Moore, "Twitter Reports 23 Million Users Are Actually 'Bots,'" PBS, August 12, 2014, accessed October 9, 2015, https://www.pbs.org/newshour/world/twitter-reports-23-million-users-actually-robot-programs.

11. Georgia Institute of Technology, "Artificial Intelligence Course Creates AI Teaching Assistant," *ScienceDaily*, May 9, 2016,

accessed May 10, 2016, http://www.sciencedaily.com/releases/2016/05/160509101930.htm.

12. Ross Miller, "AP's Robot 'Journalists' Are Writing Their Own Stories Now," *Verge*, January 29, 2015, accessed August 26, 2016, http://www.theverge.com/2015/1/29/7939067/ap-journalism-automation-robots-financial-reporting.

13. Georgia Institute of Technology, "Algorithm Allows a Computer to Create a Vacation Highlight Video: Computer Chooses Best Video for 26 Hours of Footage," *ScienceDaily*, March 10, 2016, accessed March 11, 2016, http://www.sciencedaily.com/releases/2016/03/160310125347.htm.

14. Mark Hachman, "Google's Magenta Project Just Wrote Its First Piece of Music, and Thankfully It's Not Great," *PCWorld*, June 1, 2016, accessed June 11, 2016, https://www.pcworld.com/article/415064/googles-magenta-project-just-wrote-its-first-piece-of-music-and-thankfully-its-not-great.html.

15. Hachman.

16. Associated Press, "Roombas Fill an Emotional Vacuum for Owners," NBC News, October 2, 2007, accessed November 16, 2016, https://www.nbcnews.com/id/wbna21102202.

17. Stewart E. Guthrie, "Anthropomorphism," *Encyclopaedia Britannica*, accessed December 24, 2016, http://www.britannica.com/topic/anthropomorphism.

18. Nicholas Epley et al., "On Seeing Human: A Three-Factor Theory of Anthropomorphism," *Psychological Review* 114, no. 4 (2007): 864–886.

19. Epley et al., 866.

20. Emma Seppälä, "Connectedness and Health: The Science of Social Connection," April 11, 2016, accessed December 27, 2016, https://ccare.stanford.edu/uncategorized/connectedness-health-the-science-of-social-connection-infographic/.

21. Debra Umberson and Jennifer Karas Montez, "Social Relationships and Health: A Flashpoint for Health Policy," *Journal of Health and Social Behavior* 51, suppl (2010): S54–S66, accessed December 27, 2016, https://www.ncbi.nlm.nih.gov/pmc/articles/PMC3150158/.

22. Yutaka Suzuki et al., "Measuring Empathy for Human and Robot Hand Pain Using Electroencephalography," *Scientific Reports* 5, article number 15924 (2015), doi: 10.1038/srep15924.

23. Maggie Koerth-Baker, "How Robots Can Trick You into Loving Them," *New York Times Magazine*, September 17, 2013, accessed December 13, 2016, http://www.nytimes.com/2013/09/22/magazine/how-robots-can-trick-you-into-loving-them.html.

2. OVERCOMING THE UNCANNY

1. Masahiro Mori, "The Uncanny Valley," *Energy* 7, no. 4 (1970): 34.

2. Sigmund Freud, "The Uncanny," in *The Standard Edition of the Complete Psychological Works of Sigmund Freud*, vol. 17, trans. and ed. J. Strachey (London: Hogarth Press, 1960), 219–252.

3. Joel Spolsky, "The Law of Leaky Abstractions," *Joel on Software* (blog), November 11, 2002, accessed January 23, 2017,

https://www.joelonsoftware.com/2002/11/11/the-law-of-leaky-abstractions.

4. A. P. Saygin, H. Ishiguro, J. Driver, and C. Frith, "The Thing That Should Not Be: Predictive Coding and the Uncanny Valley in Perceiving Human and Humanoid Robot Actions," *Social Cognitive and Affective Neuroscience* 7, no. 4 (2012): 413–422, doi: 10.1093/scan/nsr025.

5. Paul Clinton, "'Polar Express' a Creepy Ride," CNN, November 10, 2004, accessed January 24, 2017, http://www.cnn.com/2004/SHOWBIZ/Movies/11/10/review.polar.express/.

6. John Anderson, "Lifestyle Movie Review," *Newsday*, November 9, 2004, accessed January 24, 2017, http://www.newsday.com/lifestyles/movie-review-1.620836 (page discontinued).

7. Mary Elizabeth Williams, "Disney's 'A Christmas Carol': Bah, Humbug!," *Salon,* January 6, 2010, accessed January 24, 2017, https://www.salon.com/2009/11/06/christmas_carol_2/.

8. Manohla Dargis, "Following in Father's Parallel-Universe Footsteps," *New York Times*, December 16, 2010, accessed January 24, 2017, http://www.nytimes.com/2010/12/17/movies/17tron.html.

9. Mary Shelley, *Frankenstein; or, the Modern Prometheus*, chapter 21, page 2, Page by Page Books, accessed January 24, 2017, https://www.pagebypagebooks.com/Mary_Wollstonecraft_Shelley/Frankenstein/Chapter_21_p2.html.

10. Masahiro Mori, "The Uncanny Valley," IEEE Spectrum, June 12, 2012, accessed October 22, 2015, https://spectrum.ieee.org/the-uncanny-valley.

11. Mori.

12. Chris Weller, "The Uncanny Valley Shows How Deeply Terrified We Are of Death and Disease," Medical Daily, September 17, 2014, accessed January 6, 2017, https://www.medicaldaily.com/uncanny-valley-shows-how-deeply-terrified-we-are-death-and-disease-303568.

13. Bertrand Tondu, "Fear of the Death and Uncanny Valley: A Freudian Perspective," *Interaction Studies* 16, no. 2 (2015): 201, doi: 10.1075/is.16.2.06ton.

14. Tondu, 202.

15. Tondu, 203.

16. Kurt Gray and Daniel M. Wegner, "Feeling Robots and Human Zombies: Mind Perception and the Uncanny Valley," *Cognition* 125 (2012): 125–130, doi: 10.1016/j.cognition.2012.06.007.

17. Gray and Wegner, 129.

18. Gray and Wegner.

19. Cheyenne Laue, "Familiar and Strange: Gender, Sex, and Love in the Uncanny Valley," *Multimodal Technologies and Interaction* 1, no. 2 (2017), doi: 10.3390/mti1010002.

20. Laue.

21. University of Lincoln, "How Perfect Is Too Perfect? Research Reveals Flaws Are Key to Interacting with Humans," *ScienceDaily,* October 14, 2015, accessed March 29, 2016, https://www.sciencedaily.com/releases/2015/10/151014085142.htm.

3. COULD ROBOTS MAKE US MORE EMOTIONALLY INTELLIGENT?

1. Adrianna Hamacher et al., "Believing in BERT: Using Expressive Communication to Enhance Trust and Counteract Operational Error in Physical Human-Robot Interaction," arXiv, accessed January 17, 2017, https://arxiv.org/pdf/1605.08817v3.pdf.
2. Hamacher et al.
3. Hamacher et al.
4. Peter Salovey and John D. Mayer, "Emotional Intelligence," *Imagination, Cognition and Personality* 9, no. 3 (1990), accessed March 1, 2018, https://citeseerx.ist.psu.edu/viewdoc/download?doi=10.1.1.385.4383&rep=rep1+1type=pd.
5. Maria Popova, "The Intelligence of Emotions: Philosopher Martha Nussbaum on How Storytelling Rewires Us and Why Befriending Our Neediness Is Essential to Happiness," Marginalian, November 23, 2015, accessed December 30, 2016, https://www.themarginalian.org/2015/11/23/martha-nussbaum-upheavals-of-thought-neediness/.
6. Salovey and Mayer.
7. Megan Molteni, "The Chatbot Therapist Will See You Now," *WIRED*, June 7, 2017, accessed March 6, 2018, https://www.wired.com/2017/06/facebook-messenger-woebot-chatbot-therapist/.
8. Molteni.
9. Skye McDonald, "Will Robots Ever Have Empathy?," World Economic Forum, November 3, 2015, accessed March 2, 2018,

https://www.weforum.org/agenda/2015/11/will-robots-ever-have-empathy/.

10. McDonald.

11. Maria Popova, "What Is an Emotion? William James's Revolutionary 1884 Theory of How Our Bodies Affect Our Feelings," *Marginalian*, January 11, 2016, accessed March 8, 2018, https://www.themarginalian.org/2016/01/11/what-is-an-emotion-william-james/

12. Antonio Regalado, "What It Will Take for Computers to Be Conscious," *MIT Technology Review*, October 2, 2014, accessed March 2, 2018, https://www.technologyreview.com/2014/10/02/171077/what-it-will-take-for-computers-to-be-conscious/.

4. WILL ROBOTS BE SMARTER THAN HUMANS?

1. Katharine Child, "Afrikaans Speaking Parrot Places Amazon Order," *Times LIVE*, September 22, 2017, accessed October 24, 2017, https://www.timeslive.co.za/news/south-africa/2017-09-22-afrikaans-speaking-parrot-places-amazon-order/.

2. "Neural Networks Explained," Neuroscience News, April 17, 2017, accessed April 17, 2017, http://neurosciencenews.com/neural-networks-neuroscience-6421/.

3. Bill Steele, "Teachable Moments: Robots Learn Our Humanistic Ways," *Cornell Chronicle*, March 21, 2013, accessed August 11, 2016, https://www.news.cornell.edu/stories/2013/03/teachable-moments-robots-learn-our-humanistic-ways.

4. Michelle Starr, "Robots Learn to Cook by Watching YouTube," CNET, January 21, 2015, accessed July 12, 2016, https://www.cnet.com/au/news/robots-learning-to-cook-by-watching-youtube-videos.

5. "RoboBrain: The World's First Knowledge Engine for Robots," *MIT Technology Review*, December 12, 2104, accessed May 15, 2018, https://www.technologyreview.com/s/533471/robobrain-the-worlds-first-knowledge-engine-for-robots.

6. Amanda Schaffer, "10 Breakthrough Technologies 2016: Robots That Teach Each Other," *MIT Technology Review*, July 17, 2014, accessed October 10, 2016, https://www.technologyreview.com/s/600768/10-breakthrough-technologies-robots-that-teach-each-other/?ct=(Newsletter_2014_7_177_7_2014) (page discontinued).

7. John Bohannon, "A New Breed of Scientist, with Brains of Silicon," *Science*, July 5, 2017, accessed September 14, 2017, http://www.sciencemag.org/news/2017/07/new-breed-scientist-brains-silicon.

8. Jason Tanz, "Soon We Won't Program Computers. We'll Train Them Like Dogs," *WIRED*, May 17, 2016, accessed July 11, 2016, http://www.wired.com/2016/05/the-end-of-code/.

9. Tanz.

10. Will Knight, "The Dark Secret at the Heart of AI," *MIT Technology Review*, April 11, 2017, accessed October 5, 2017, https://www.technologyreview.com/s/604087/the-dark-secret-at-the-heart-of-ai/.

11. Aaron Frank, "Why the Death of Moore's Law Could Give Birth to More Human-Like Machines," *WIRED*, August 10, 2016,

accessed September 9, 2016, https://www.wired.co.uk/article/moores-law-ending-good-ai.

12. Will Knight, "AI Could Get 100 Times More Energy-Efficient with IBM's New Artificial Synapses," *MIT Technology Review*, June 12, 2018, accessed June 12, 2018, https://www.technologyreview.com/2018/06/12/142361/ai-could-get-100-times-more-energy-efficient-with-ibms-new-artificial-synapses/.

13. "The Rise in Computing Power: Why Ubiquitous Artificial Intelligence Is Now a Reality," *Forbes*, July 17, 2018, accessed July 24, 2018, https://www.forbes.com/sites/intelai/2018/07/17/the-rise-in-computing-power-why-ubiquitous-artificial-intelligence-is-now-a-reality/.

14. V. C. Müller and Nick Bostrom, "Future Progress in Artificial Intelligence: A Survey of Expert Opinion," in *Fundamental Issues of Artificial Intelligence*, ed. V. C. Müller (Cham, Switzerland: Springer, 2016), 555–572.

15. David J. Chalmers, "The Singularity: A Philosophical Analysis," accessed July 31, 2018, https://consc.net/papers/singularity.pdf.

5. DO ROBOTS SPELL DOOMSDAY FOR THE HUMAN RACE?

1. Matthew Graves, "Why We Should Be Concerned About Artificial Superintelligence," *Skeptic*, accessed July 30, 2018, https://www.skeptic.com/reading_room/why-we-should-be-concerned-about-artificial-superintelligence/.

2. Cade Metz and Gregory Schmidt, "Elon Musk and Others Call for Pause on A.I., Citing 'Profound' Risks to Society," *New York Times*, March 29, 2023, accessed March 29, 2023, https://www.nytimes.com/2023/03/29/technology/ai-artificial-intelligence-musk-risks.html.

3. Stephen Hawking, Stuart Russell, Max Tegmark, and Frank Wilczek, "Stephen Hawking: 'Transcendence Looks at the Implications of Artificial Intelligence—But Are We Taking AI Seriously Enough?,'" *Independent*, May 1, 2014, accessed July 31, 2018, https://www.independent.co.uk/news/science/stephen-hawking-transcendence-looks-at-the-implications-of-artificial-intelligence-but-are-we-taking-9313474.html.

4. Hawking et al.

5. John Markoff, "Scientists Worry Machines May Outsmart Man," *New York Times*, July 25, 2009, accessed July 31, 2018, https://www.nytimes.com/2009/07/26/science/26robot.html.

6. Doug Bolton, "'Artificial Intelligence Alarmists' like Elon Musk and Stephen Hawking Win 'Luddite of the Year Award,'" *Independent*, January 19, 2016, accessed July 31, 2018, https://www.independent.co.uk/life-style/gadgets-and-tech/news/elon-musk-stephen-hawking-luddite-award-of-the-year-itif-a6821921.html.

7. Kate Baggaley, "There Are Two Kinds of AI, and the Difference Is Important," *Popular Science*, February 23, 2017, accessed May 29, 2018, https://www.popsci.com/narrow-and-general-ai.

8. John R. Searle, "Minds, Brains, and Programs," *Behavioral and Brain Sciences* 3, no. 3 (1980): 417–424, accessed August 13,

2018, https://www.cambridge.org/core/journals/behavioral-and-brain-sciences/article/abs/minds-brains-and-programs/DC644B47A4299C637C89772FACC2706A.

9. Searle.

10. Steven Pinker, "The Dangers of Worrying About Doomsday," *Globe and Mail*, February 26, 2018, accessed July 31, 2018, http://www.theglobeandmail.com/opinion/the-dangers-of-worrying-about-doomsday/article38062215/.

11. Melissa Schilling, "Elon Musk Fires Back at Harvard Psychologist Steven Pinker Over the Future of Artificial Intelligence," *Inc.*, March 2, 2018, accessed July 31, 2018, https://www.inc.com/melissa-schilling/how-to-make-sense-of-clash-between-elon-musk-stephen-pinker-over-artificial-intelligence.html.

12. Catherine Clifford, "Harvard Psychologist Steven Pinker: The Idea That A.I. Will Lead to the End of Humanity Is Like the Y2K Bug" February 27, 2018, accessed August 14, 2023, https://www.cnbc.com/2018/02/27/harvard-psychologist-steven-pinker-on-artificial-intelligence.html.

13. Clifford.

14. David J. Chalmers, "The Singularity: A Philosophical Analysis," accessed July 31, 2018, https://consc.net/papers/singularity.pdf.

15. James Vincent, "Twitter Taught Microsoft's AI Chatbot to Be a Racist Asshole in Less Than a Day," Verge, March 24, 2016, accessed August 22, 2018, https://www.theverge.com/2016/3/24/11297050/tay-microsoft-chatbot-racist.

6. LONELINESS CAN KILL YOU. COULD A ROBOT SAVE YOUR LIFE?

1. Claire Pomeroy, "Loneliness Is Harmful to Our Nation's Health: Research Underscores the Role of Social Isolation in Disease and Mortality," *Scientific American*, March 20, 2019, accessed November 6, 2019, https://blogs.scientificamerican.com/observations/loneliness-is-harmful-to-our-nations-health.

2. Ellie Polak, "New Cigna Study Reveals Loneliness at Epidemic Levels in America," Cigna, May 1, 2018, accessed November 6, 2019, https://www.multivu.com/players/English/8294451-cigna-us-loneliness-survey/.

3. Pomeroy.

4. Brad Porter, "Loneliness Might Be a Bigger Health Risk Than Smoking or Obesity," *Forbes*, January 18, 2017, accessed October 12, 2017, https://www.forbes.com/sites/quora/2017/01/18/loneliness-might-be-a-bigger-health-risk-than-smoking-or-obesity/?sh=7cb459f25d13.

5. Porter.

6. Katharine Gammon, "Why Loneliness Can Be Deadly," Live Science, March 2, 2012, accessed October 12, 2017, https://www.livescience.com/18800-loneliness-health-problems.html.

7. Judith Shulevitz, "The Lethality of Loneliness: We Now Know How It Can Ravage Our Body and Brain," *New Republic*, May 13, 2013, accessed October 12, 2017, https://newrepublic.com/article/113176/science-loneliness-how-isolation-can-kill-you.

8. Alex Hacillo, "Lonely in Tokyo," Medium, November 10, 2015, accessed September 7, 2017, https://medium.com/the-megacities-issue/lonely-in-tokyo-e4d0b89c17f.

9. Mizuho Aoki, "In Sexless Japan, Almost Half of Single Young Men and Women Are Virgins: Survey," *Japan Times*, September 9, 2016, accessed November 7, 2019, https://www.japantimes.co.jp/news/2016/09/16/national/social-issues/sexless-japan-almost-half-young-men-women-virgins-survey/.

10. Philippe Mesmer, "Rent-A-Friend: A Solution for the Lonely People of Japan," World Crunch, January 15, 2014, accessed September 17, 2017, https://worldcrunch.com/culture-society/rent-a-friend-a-solution-for-the-lonely-people-of-japan.

11. Rachel Lowry, "Meet the Lonely Japanese Men in Love with Virtual Girlfriends," *Time*, September 15, 2015, accessed September 7, 2017, http://www.time.com/3998563/virtual-love-japan/.

12. Andrew McCormick, "Asia's Lonely Youth Are Turning to Machines for Companionship and Support," *South China Morning Post*, June 16, 2018, accessed June 19, 2018, https://www.scmp.com/tech/article/2150720/asias-lonely-youth-are-turning-machines-companionship-and-support.

13. Nick Charity, "Japanese Man Marries Hologram He Admired for Ten Years in Tokyo Ceremony," *Standard*, November 15, 2018, accessed January 20, 2020, https://www.standard.co.uk/news/world/japanese-man-marries-the-hologram-he-admired-for-10-years-in-tokyo-ceremony/-a3991401.html.

14. Charity.

15. Thisanka Siripala, "Japan's Robot Revolution in Senior Care," *Diplomat*, June 1, 2018, accessed November 4, 2019, https://www.thediplomat.com/2018/06/japans-robot-revolution-in-senior-care/.

16. Malcolm Foster, "Aging Japan: Robots May Have Role in Future of Elder Care," Reuters, March 27, 2018, accessed July 8, 2019, https://www.reuters.com/article/us-japan-ageing-robots-widerimage/aging-japan-robots-may-have-role-in-future-of-elder-care-idUSKBN1H33AB .

17. Siripala.

18. Foster.

19. "Over 80% of Japanese Would Welcome Robot Caregivers," Nippon.com, December 4, 2018, accessed November 4, 2019, https:/www.nippon.com/en/features/h00342/.

20. Sophie Knight, "Japan's Irresistible Cult of Cuteness," *Post-Gazette*, July 24, 2016, accessed September 5, 2017, http://www.post-gazette.com/opinion/Op-Ed/2016/07/24/Japan-s-irresistible-cult-of-cuteness/stories/201607240039.

21. Marc Prosser, "Japan Loves Robots. Japan Loves Cute Things. The Combination of the Two? Unmissable," RedBull.com, March 22, 2017, accessed September 5, 2017, https://www.redbull.com/us-en/japan-cute-robot-obsession (page discontinued).

22. Kevin Lynch, "Robot Astronaut Kirobo Sets Two Guinness World Records Titles," Guinness World Records, March 27, 2015, accessed November 30, 2019, https://www.guinnessworldrecords.com/news/2015/3/robot-astronaut-kirobo-sets-two-guinness-world-records-titles-375259.

23. Raffaele Rodogno, "Social Robots, Fiction, and Sentimentality," *Ethics and Information Technology* 18, no. 4 (August 2015), accessed July 15, 2019, https://pure.au.dk/portal/files/90856923/Social_Robots_and_Sentimentality_Pre_Final.pdf.

24. Rodogno.

7. LOVE IN THE TIME OF ROBOTS

1. James Cook, "Sex Robots in 2022—What's Happened in the Last Four Years?" Business Leader, March 21, 2022, https://www.businessleader.co.uk/sex-robots-in-2022-whats-happened-in-the-last-four-years/#~text=In 2022%2C.

2. Sven Nyholm and Lily Eva Frank, "From Sex Robots to Love Robots: Is Mutual Love with a Robot Possible?," MIT Press Scholarship Online, accessed March 28, 2022, doi: 10.7551/mitpress/9780262036689.003.0012.

3. Sherry Turkle, *Alone Together: Why We Expect More from Technology and Less from Each Other* (New York: Basic Books, 2011), 19.

4. Nyholm and Frank.

5. Charles Q. Choi, "Humans Marrying Robots? A Q&A with David Levy," *Scientific American*, February 19, 2008, accessed November 22, 2015, http://www.scientificamerican.com/article/humans-marrying-robots/.

6. David Levy, *Love and Sex with Robots: The Evolution of Human-Robot Relationships* (New York: HarperCollins, 2007), 194–195.

7. Jessica Masterson, "More Women May Be Paying for Sex, but This Does Not Equate to Female Sexual Liberation," Feminist Current, November 12, 2019, accessed April 11, 2022, https://www.feministcurrent.com/2019/11/12/more-women-may-be-paying-for-sex-but-this-does-not-signal-female-sexual-liberation/.

8. Danielle Knafo, "Guys and Dolls: Relational Life in the Technological Era," *Psychoanalytic Dialogues*, July 2015, accessed August 31, 2016, https://www.researchgate.net/publication/282520564_Guys_and_Dolls_Relational_Life_in_the_Technological_Era.

9. Knafo.

10. Holly Ellyatt, "Campaign Launched Against 'Harmful' Sex Robots," CNBC, September 15, 2015, accessed March 31, 2018, http://www.cnbc.com/2015/09/15/sex-robots-campaign.html.

11. Justin Moyer, "Having Sex with Robots Is Really, Really Bad, Campaign Against Sex Robots Says," *Washington Post*, September 15, 2015, accessed March 3, 2022, https://www.washingtonpost.com/news/morning-mix/wp/2015/09/15/having-sex-with-robots-is-really-really-bad-campaign-against-sex-robots-says/.

8. IS THERE A ROBOT NANNY IN YOUR CHILDREN'S FUTURE?

1. "Child Care Costs by State 2022," World Population Review, accessed May 9, 2022, https:/www.worldpopulationreview.com/state-rankings/child-care-costs-by-state.

2. Alice LaPlante, "Robot Nannies Are Here, but Won't Replace Your Babysitter—Yet," *Forbes*, March 29, 2017, accessed November

9, 2018, https://www.forbes.com/sites/centurylink/2017/03/29/robot-nannies-are-here-but-wont-replace-your-babysitter-yet/?sh=58d2c31256b7.

3. Daniel Xiong, email interview with the author, December 20, 2018.
4. Xiong.
5. Claire A. G. J. Huijnen et al., "Matching Robot KASPAR to Autism Spectrum Disorder (ASD) Therapy and Educational Goals," *International Journal of Social Robotics* 8 (2016): 445–455, accessed August 11, 2016, https://link.springer.com/article/10.1007/s12369-016-0369-4.
6. "Why Do Children with Autism Learn Better from Robots?," LuxAI, accessed June 1, 2022, https://luxai.com/blog/why-children-with-autism-learn-better-from-robots/.
7. William Weir, "Robots Help Children with Autism Improve Social Skills," Yale News, August 22, 2018, accessed June 1, 2022, https://news.yale.edu/2018/08/22/robots-help-children-autism-improve-social-skills.
8. Bosede I. Edwards and Adrian D. Cheok, "Why Not Robot Teachers: Artificial Intelligence for Addressing Teacher Shortage," *Applied Artificial Intelligence* 32, no. 4 (2018): 345–360, https://doi.org/10.1080/08839514.2018.1464286.
9. James Manyika et al., *A Future That Works: Automation, Employment, and Productivity* (n.p.: McKinsey Global Institute, January 2017).
10. Sherry Turkle, "Why These Friendly Robots Can't Be Good Friends to Our Kids," *Washington Post*, December 7, 2017, ac-

cessed November 13, 2018, https://www.washingtonpost.com/outlook/why-these-friendly-robots-cant-be-good-friends-to-our-kids/2017/12/07/bce1eaea-d54f-11e7-b62d-d9345ced896d_story.html.

11. "Electronic Baby Toys Associated with Decrease in Quality and Quantity of Language in Infants," Neuroscience News, December 31, 2015, accessed November 26, 2018, https://neurosciencenews.com/toys-language-neurodevelopment-3330.

12. Noel Sharkey and Amanda Sharkey, "The Crying Shame of Robot Nannies," *Interaction Studies* 11, no. 2 (2010): 161–190, doi: 10.1075/is.11.2.01sha.

9. KILLING MACHINES OR COMBAT BUDDIES?

1. "The Quiet Professional: An Investigation of U.S. Military Explosive Ordnance Disposal Personnel Interactions with Everyday Field Robots," https://digital.lib.washington.edu/researchworks/handle/1773/24197?show=full

2. Fox Van Allen, "The Deadly, Incredible and Absurd Robots of the U.S. Military," CNET, February 18, 2017, accessed November 23, 2022, https://www.cnet.com/pictures/deadly-incredible-absurd-robots-the-us-military/.

3. Kyle Chayka, "As Military Robots Increase, So Does the Complexity of Their Relationship with Soldiers," *Newsweek*, February 18, 2014, accessed November 14, 2022, https://www.newsweek.com/2014/02/21/military-robots-increase-so-does-complexity-their-relationship-soldiers-245530.html.

4. Marijn Hoijtink and Marlene Tröstl, "The Intimacies of Soldier-Robot Relations," Utrecht University website, *Intimacies of Remote Warfare* podcast, May 31, 2021, accessed March 23, 2023, https://intimacies-of-remote-warfare.nl/podcasts-documentaries/the-intimacies-of-soldier-robot-relations-and-the-making-of-remote-warfare/.
5. Chayka.
6. Chayka.
7. Kostantin Toropin, "Marine Corps Planning for Wars Where Robots Kill Each Other," Military.com, September 15, 2022, accessed November 8, 2022, https://www.military.com/daily-news/2022/09/15/marine-corps-planning-wars-where-robots-kill-each-other.html.
8. Patrick Lin et al., "Robots in War: Issues of Risk and Ethics," in *Ethics and Robotics*, eds. R. Capurro and M. Nagenborg (Amsterdam: IOS Press, 2009).
9. Sam Thielman, "Use of Police Robot to Kill Dallas Shooting Suspect Believed to Be First in U.S. History," *Guardian*, July 8, 2016, accessed November 30, 2022, https://www.theguardian.com/technology/2016/jul/08/police-bomb-robot-explosive-killed-suspect-dallas.

10. HOW WILL ROBOTS CHANGE HUMAN CULTURE?

1. https://moxierobot.com/?gclid=CjwKCAjw7p6aBhBiEiwA83fGuoDYDG8GNNmqmsCtN28rCZ3c7VDYNY0yyszaBgQuLyxd_mNAMkC7qxoCsFoQAvDBwE.

2. Patrick Lucas Austin, "Sony's New Aibo Robot Dog Is Nearly $3,000, but It Can Tell How You're Feeling," *Time*, August 23, 2018, accessed October 6, 2022, https://www.yahoo.com/news/sony-apos-aibo-robot-dog-204113573.html.

3. Peter H. Kahn Jr. et al., "Children's Social Relationships with Current and Near-Future Robots," *Child Development Perspectives* 7, no. 1 (2013): 32–37.

4. https://pubmed.ncbi.nlm.nih.gov/22369338/, Accessed August 14, 2023

5. https://www.academia.edu/14911043/Will_humans_mutually_deliberate_with_social_robots, Accessed August 14, 2023.

6. https://sites.dartmouth.edu/dujs/2018/10/15/robotic-peer-pressure-how-robots-can-influence-childrens-opinions/, Accessed August 14, 2023.

7. Jacqueline M. Kory-Westlund and Cynthia Breazeal, "A Long-Term Study of Young Children's Rapport, Social Emulation, and Language Learning with a Peer-Like Robot Playmate in Preschool," *Frontiers in Robots and AI* 6 (2019), accessed October 20, 2022, https://doi.org/10.3389/frobt.2019.00081.

8. "The History of Electricity | History of Electricity Timeline," Mr. Electric, accessed October 27, 2022, https://mrelectric.com/blog/the-history-of-electricity-history-of-electricity-timeline.

9. "A (Mostly) Quick History of Smartphones," Cellular Sales, September 28, 2021, accessed October 27, 2022, https://www.cellularsales.com/blog/a-mostly-quick-history-of-smartphones/.

10. Matthew J. Taylor and Candace A. Thoth, "Cultural Transmission," in *Encyclopedia of Child Behavior and Development*,

eds. Jack Naglieri and Sam Goldstein (Boston: Springer, 2011), 448.

11. Xin Qin et al., “Adults Still Can’t Resist: A Social Robot Can Induce Normative Conformity,” *Computers in Human Behavior* 129 (April 2022), accessed October 27, 2022, https://doi.org/10.1016/j.chb.2021.107041.
12. Qin et al.
13. Michele Marenus, “Howard Gardner’s Theory of Multiple Intelligences,” *Simply Psychology*, June 9, 2020, accessed October 25, 2022, https://www.simplypsychology.org/multiple-intelligences.html.
14. Marenus.

11. THE GOOD NEWS: HUMANS ARE IN CONTROL. THE BAD NEWS: HUMANS ARE IN CONTROL.

1. Bernard Siegel, interview with the author, July 8, 2022.
2. Mark Coeckelbergh, “Robot Rights? Towards a Social-Relational Justification of Moral Consideration,” *Ethics of Information Technology* 12 (2010): 209–221, doi: 10.1007/s10676-010-9235–5.
3. Coeckelbergh.
4. Jane Wakefield, “Are You Scared Yet? Meet Norman, the Psychopathic AI,” BBC, June 2, 2018, accessed June 5, 2022, https://www.bbc.com/news/technology-44040008.